Unit 3
Surviving in Changing Environments

Copyright © 2020 by Discovery Education, Inc. All rights reserved. No part of this work may be reproduced, distributed, or transmitted in any form or by any means, or stored in a retrieval or database system, without the prior written permission of Discovery Education, Inc.

NGSS is a registered trademark of Achieve. Neither Achieve nor the lead states and partners that developed the Next Generation Science Standards were involved in the production of this product, and do not endorse it.

To obtain permission(s) or for inquiries, submit a request to:

Discovery Education, Inc.
4350 Congress Street, Suite 700
Charlotte, NC 28209
800-323-9084
Education_Info@DiscoveryEd.com

ISBN 13: 978-1-68220-796-3

Printed in the United States of America.

1 2 3 4 5 LBC 28 27 26 25 24   A

Acknowledgments

Acknowledgment is given to photographers, artists, and agents for permission to feature their copyrighted material.

Cover and inside cover art: Vladimir Wrangel / Shutterstock.com

Table of Contents

Unit 3: Surviving in Changing Environments

Letter to the Parent/Guardian vi

Unit Overview vii

   Anchor Phenomenon: Who Left the Tracks? 2

   Unit Project Preview: Environmental Changes and Animals 4

Concept 3.1 Understanding Fossils

Concept Overview 6

   Wonder 8

      Investigative Phenomenon: Fossils 10

   Learn 22

   Share 46

Concept 3.2 Interactions in the Environment

Concept Overview 54

   Wonder 56

      Investigative Phenomenon: The Kangaroo Rat 58

   Learn 74

   Share 126

Concept 3.3 Environmental Changes

Concept Overview 136

    Wonder.. 138

        Investigative Phenomenon: The Salt Harvest Mouse 140

    Learn... 148

    Share... 186

Unit Wrap-Up

Unit Project: Environmental Changes and Animals............... 196

Grade 3 Resources

Bubble Map R3

Safety in the Science Classroom R4

Vocabulary Flash Cards R7

Glossary R27

Index R52

Dear Parent/Guardian,

This year, your student will be using Science Techbook™, a comprehensive science program developed by the educators and designers at Discovery Education and written to the Next Generation Science Standards (NGSS). The NGSS expect students to act and think like scientists and engineers, to ask questions about the world around them, and to solve real-world problems through the application of critical thinking across the domains of science (Life Science, Earth and Space Science, Physical Science).

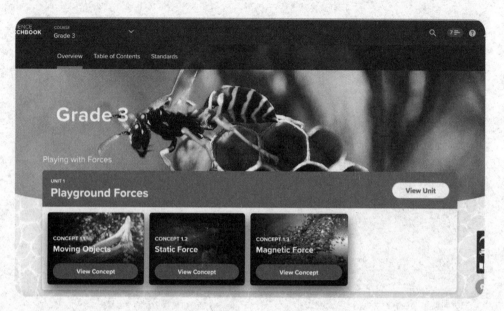

Science Techbook is an innovative program that helps your student master key scientific concepts. Students engage with interactive science materials to analyze and interpret data, think critically, solve problems, and make connections across science disciplines. Science Techbook includes dynamic content, videos, digital tools, Hands-On Activities and labs, and game-like activities that inspire and motivate scientific learning and curiosity.

You and your child can access the resource by signing in to www.discoveryeducation.com. You can view your child's progress in the course by selecting the Assignment button.

Science Techbook is divided into units, and each unit is divided into concepts. Each concept has three sections: Wonder, Learn, and Share.

Units and Concepts Students begin to consider the connections across fields of science to understand, analyze, and describe real-world phenomena.

Wonder Students activate their prior knowledge of a concept's essential ideas and begin making connections to a real-world phenomenon and the **Can You Explain?** question.

Learn Students dive deeper into how real-world science phenomenon works through critical reading of the Core Interactive Text. Students also build their learning through Hands-On Activities and interactives focused on the learning goals.

Share Students share their learning with their teacher and classmates using evidence they have gathered and analyzed during Learn. Students connect their learning with STEM careers and problem-solving skills.

Within this Student Edition, you'll find QR codes and quick codes that take you and your student to a corresponding section of Science Techbook online. To use the QR codes, you'll need to download a free QR reader. Readers are available for phones, tablets, laptops, desktops, and other devices. Most use the device's camera, but there are some that scan documents that are on your screen.

For resources in Science Techbook, you'll need to sign in with your student's username and password the first time you access a QR code. After that, you won't need to sign in again, unless you log out or remain inactive for too long.

We encourage you to support your student in using the print and online interactive materials in Science Techbook, on any device. Together, may you and your student enjoy a fantastic year of science!

Sincerely,

The Discovery Education Science Team

Unit 3
Surviving in Changing Environments

Get Started

Who Left the Tracks?

Have you ever seen tracks in the mud or the snow? How did you know what animal made the tracks? Do you think that hundreds of years ago animal tracks looked the same as they do today? In this unit, you will explore how fossils give us clues about the types of animals that lived in the past and where they lived. You will also investigate how changes to the environment cause changes in the traits of living organisms.

Quick Code: us3506s

Who Left the Tracks?

Think About It

Look at the photograph. **Think** about the following questions.

- How does the environment affect living organisms?
- How do organisms' traits help them survive in different environments?
- What happens to organisms when the environment changes?

Tracks

Unit 3: Surviving in Changing Environments | 3

Unit Project Preview

Solve Problems Like a Scientist

Unit Project: Environmental Changes and Animals

In this project, you will use what you know about changes in the environment and in organisms' traits to create a story of the geologic history for your area.

Quick Code: us3507s

Paleontologist

SEP Planning and Carrying Out Investigations
CCC Stability and Change

Ask Questions About the Problem

You are going to create a geologic history of your area using what you know about how changing environments affect organisms. **Write** some questions you can ask to learn more about the problem. As you learn about how organisms survive in changing environments in this unit, **write down** the answers to your questions.

Unit 3: Surviving in Changing Environments

CONCEPT 3.1

Understanding Fossils

Student Objectives

By the end of this lesson:

- [ ] I can show evidence that extinct organisms are related to living organisms.
- [ ] I can describe how a fossil provides information about ancient environments.
- [ ] I can use evidence to figure out where a fossil would be found.

Key Vocabulary

- [ ] ancient
- [ ] Arctic
- [ ] dinosaur
- [ ] environment
- [ ] extinct
- [ ] fossil
- [ ] microorganism
- [ ] organism
- [ ] prehistoric
- [ ] species
- [ ] trait
- [ ] tropical

Quick Code: us3505s

Concept 3.1: Understanding Fossils | 7

Activity 1
Can You Explain?

How can fossils help us understand ancient environments?

_____
_____
_____
_____
_____
_____
_____
_____

Quick Code:
us3509s

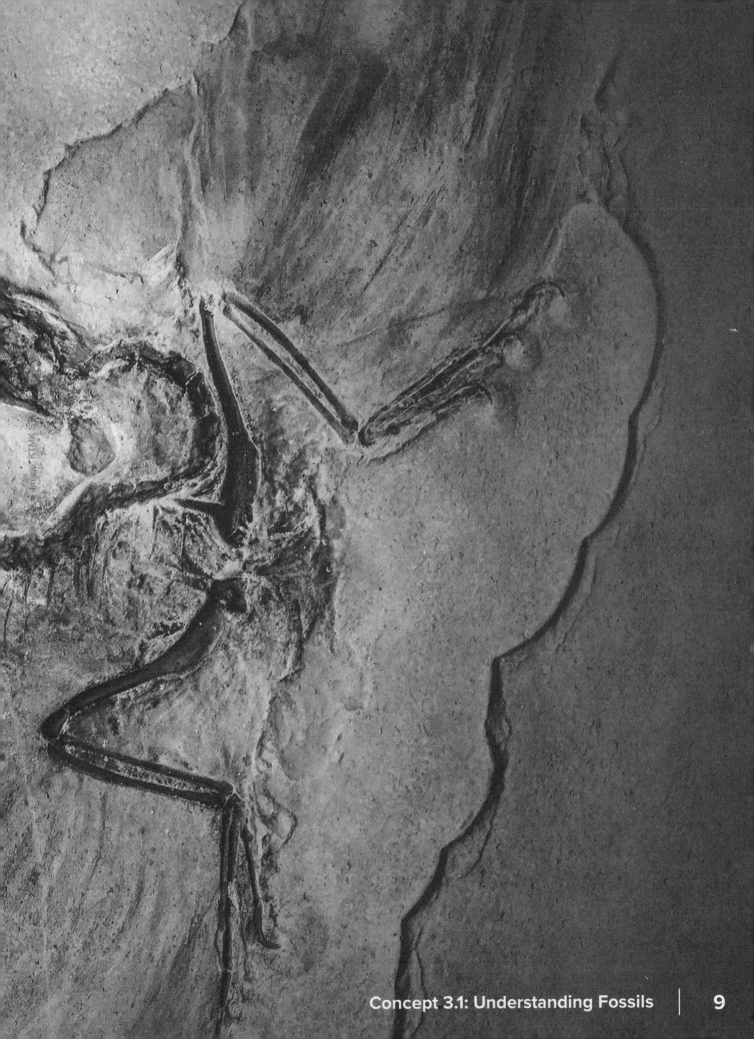

Concept 3.1: Understanding Fossils | 9

3.1 | Wonder — How can fossils help us understand ancient environments?

Activity 2

Ask Questions Like a Scientist

Fossils

Quick Code: us3510s

Look at the image. Then, **write** your answers to the questions.

Let's Investigate Fossils

SEP Asking Questions and Defining Problems

Is this picture something old or something new? Why do you think so?

_____

_____

_____

What do you think made this shape? Why do you think so?

_____

_____

_____

Write a question you have about the image that you would like to answer.

_____

_____

_____

Concept 3.1: Understanding Fossils

Messages in Fossils

Today, we have a good idea of what the planet looked like long ago and what plants and animals were alive millions of years ago. Scientists like paleontologists have found answers by studying **fossils.**

Although fossils may look just like rocks initially, fossils are important **evidence** that can teach us about **prehistoric** life on Earth. Paleontologists look for clues in fossils to learn about life and **environments** long ago. Paleontologists can search for **dinosaur** fossils to learn about life when dinosaurs roamed Earth!

Dinosaur Footprint

Have you ever wondered what the world looked like when dinosaurs were alive? How did they move? What did they eat? Where did they live? When you study a fossil, you can find the answers to your questions about the world long ago.

Activity 3
Analyze Like a Scientist

Quick Code: us3511s

Messages in Fossils

Read the text. **Look** at the image. Then, **write** your ideas about the questions you read in the text.

My Ideas:

_____

_____

_____

_____

_____

_____

_____

_____

_____

SEP Analyzing and Interpreting Data

Concept 3.1: Understanding Fossils | 13

3.1 | Wonder
How can fossils help us understand ancient environments?

Activity 4
Observe Like a Scientist

Fossils and Habitats

Look at the image. Then, **answer** the question.

Quick Code: us3512s

Dinosaur on the Run

SEP Analyzing and Interpreting Data
CCC Scale, Proportion, and Quantity

How were scientists able to predict what the dinosaur and its environment looked like?

3.1 | Wonder How can fossils help us understand ancient environments?

Watch the video segments. As you watch, **complete** the vocabulary chart about the word *fossil*:

Fossils

Fossil Formation

Definition
- Personal:

- Dictionary:

Examples (drawn or written)

Fossil

Sentences
- Teacher/Book:

- Personal:

Related:

Word Parts:

Outside of School (Who would use the word? How would they use it?):

Concept 3.1: Understanding Fossils | 17

3.1 | Wonder How can fossils help us understand ancient environments?

Activity 5
Evaluate Like a Scientist

Quick Code: us3513s

What Do You Already Know About Understanding Fossils?

What Is a Fossil?

Look at each of the images. **Write** "fossil" next to the images that are fossils and "not a fossil" next to the images that are not fossils.

_____

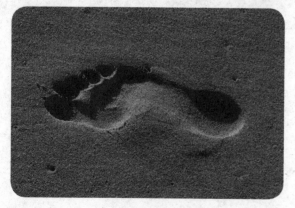

_____

SEP Analyzing and Interpreting Data
CCC Scale, Proportion, and Quantity

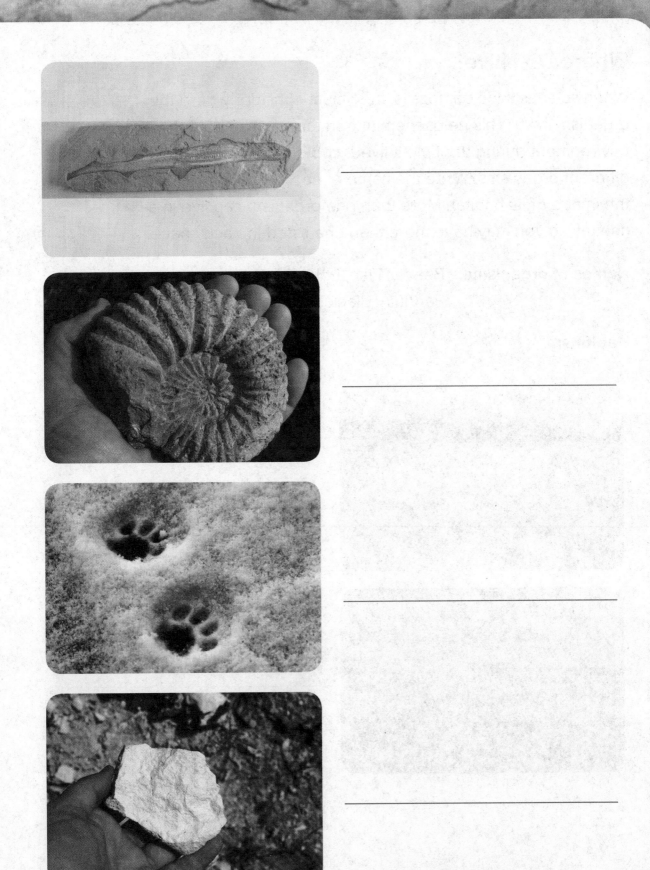

Concept 3.1: Understanding Fossils

3.1 | Wonder — How can fossils help us understand ancient environments?

Where Do I Live?

When scientists look at fossils, they must consider where the organism lived. This helps them understand more about the environment during the time it lived. **Look** at the names of different organisms. **Write** the name of each organism next to the image of its habitat. More than one organism can live in a habitat, so you may have more than one listed in each space.

Names of organisms: Beaver, Fish, Polar Bear, Antelope, Penguin, Black Bear

Habitats:

_____

_____

What Is a Fossil?

Circle each statement that correctly describes a fossil:

A. A fossil can be made by an impression left by a living thing.

B. A fossil can be created when a living thing dies.

C. A fossil can be evidence for change over time.

D. A fossil can only be made from old living things.

E. A fossil can be made from the leftover hard parts of a living thing.

F. A fossil can be made in a few years.

Concept 3.1: Understanding Fossils | 21

What Are Fossils?

Activity 6
Analyze Like a Scientist

What Are Fossils?

Quick Code: us3514s

Fossils give evidence of change over time. **Read** the text. As you read, **highlight** information you can use as evidence to support the claim that fossils give evidence of change over time.

What Are Fossils?

Fossils are evidence of **ancient** life. They are the preserved remains of **organisms** that lived thousands, or even millions, of years ago.

Fossils can be formed from plants, animals, and even **microorganisms**. Fossils can look very much like the original object. You may have seen fossilized bones in a museum or fossilized shells in a rock. You may have even found a beautiful piece of petrified wood. In these objects, the original bone, shell, or wood has been replaced over time by rock. Fossils can also be imprints in rock. These include things like dinosaur tracks, human footprints, or even paths made by slithering snakes.

SEP **Engaging in Argument from Evidence**

Activity 7
Observe Like a Scientist

Quick Code: us3515s

Earth History

Complete the Forming Fossils part of the interactive Earth History to learn about how fossils are formed. Then, **record** the steps necessary to make a fossil in your Summary Frames graphic organizer.

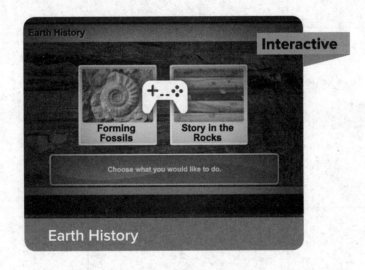

Concept 3.1: Understanding Fossils

3.1 | Learn
How can fossils help us understand ancient environments?

Topic: _____

_____   _____   _____

_____   _____   _____

_____   _____   _____

How Can You Find Clues in Fossils?

Activity 8
Analyze Like a Scientist

Quick Code:
us3516s

Clues in Fossils

Read the text and **highlight** key information that describes how fossils are evidence of changes in organisms over time.

Clues in Fossils

Fossils can be of organisms that are now **extinct**. But some fossils may be related to an organism that exists today. By studying a fossil, you can discover how an organism may have changed over time to adapt to new, changing environments.

Where did this fossil live?

Clues exist in every fossil. How large is the organism? What features does the organism have? How might those features have interacted with the environment?

SEP Analyzing and Interpreting Data

Concept 3.1: Understanding Fossils | 25

Analyze the images. Then, **complete** the sentences for each fossil.

Can you see features of this ancient animal that are similar to something you see today?

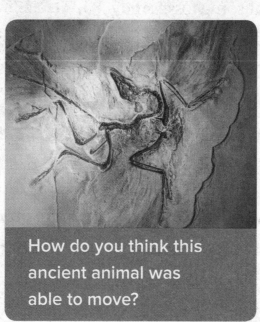

How do you think this ancient animal was able to move?

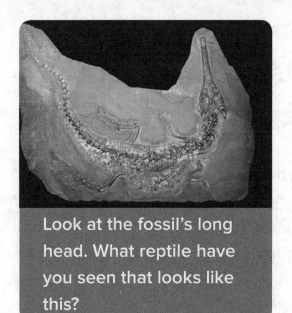

Look at the fossil's long head. What reptile have you seen that looks like this?

The fossil reminds me of a(n) _____ because of its _____.

The fossil reminds me of a(n) _____ because of its _____.

The fossil reminds me of a(n) _____ because of its _____.

Where do you think these organisms would live if they were alive today?

_____

_____

Concept 3.1: Understanding Fossils

3.1 | Learn How can fossils help us understand ancient environments?

Activity 9

Think Like a Scientist

Where Did I Come From?

Quick Code: us3517s

In this investigation, you will **analyze** images of fossils. You will make a claim about what the environment was like where that fossil existed. You will also **explain** how the environment where the fossil lived is different from the environment where it was found.

What materials do you need? (per group)

- Fossil cards
- Fossil habitat description cards
- Location description cards

What Will You Do?

1. Analyze the images of fossils and the list of habitats provided by your teacher.

2. Make predictions about what the environment was like where and when the fossil was a living organism.

3. Match the fossil with the habitat that is from when the fossil was a living organism.

4. Select two images. Write a claim that describes the environment in which the fossil lived.

SEP Analyzing and Interpreting Data

Look at each fossil picture. **Read** about the habitats. Which fossil would have come from each habitat? **Write** the name of the habitat next to the picture of the fossil.

Forest Habitat: I lived in a place where there was a good source of water to help me grow, but I did not live in the water. I lived in a place where the temperature was not too hot or too cold. Different types of animals would use me for shelter.

Underwater Habitat: I lived underwater. I could feel warmth from the sun, but I could not actually see the sun. It was mostly warm where I lived. I lived with other animals such as fish and plankton.

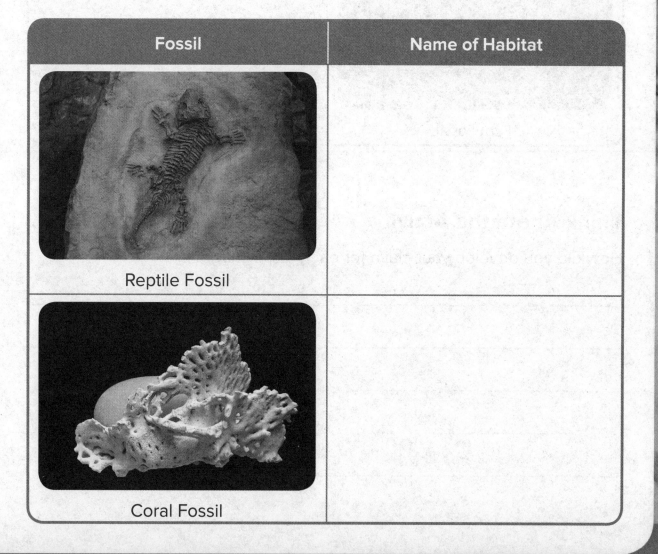

Fossil	Name of Habitat
Reptile Fossil	
Coral Fossil	

Concept 3.1: Understanding Fossils

3.1 | Learn   How can fossils help us understand ancient environments?

Petrified Tree Fossil

Fern Fossil

Think About the Activity

How did you develop your claim for each fossil image?

_____

_____

_____

_____

What similarities and differences did you notice about the environment where each prehistoric animal and plant lived?

_____
_____
_____
_____
_____

Do you think the environment where the fossil was found looks the same as it did millions of years ago?

_____
_____
_____
_____
_____
_____
_____

3.1 | Learn — How can fossils help us understand ancient environments?

Activity 10

Investigate Like a Scientist

Quick Code: us3518s

Hands-On Investigation: Making Fossils

In this activity, you will work with a group of other students to **make** different types of fossils. Then, you will **share** your fossils with a different group and the new group must **make observations** and **predict** what the fossil is and what its environment must have looked like.

Make a Prediction

What kinds of plants and animals do you think will make the best fossils?

_____

_____

_____

_____

What Will You Do?

1. Go outside and look for an object that you can fossilize. The object should be a living thing, like a leaf or an acorn.

2. Mash a ball of fossil dough until it looks like a flat disc and then press the object into the dough until it forms an imprint. Set the dough aside to let the pieces dry and turn into "fossils."

3. Share your dried fossil with another group. Use hand lenses to examine the fossil that was shared with you. Use the ruler to make measurements of the fossil.

What materials do you need? (per group)

- Outside items to fossilize (twig, leaf, acorn, etc.)
- Wax paper
- Metric ruler
- Hand lens
- Cornstarch
- Baking soda (sodium bicarbonate)
- $2\frac{1}{2}$ cups cold water

Think About the Activity

How were you able to predict what the fossil was?

_____
_____
_____
_____
_____
_____
_____

SEP Analyzing and Interpreting Data

3.1 | Learn — How can fossils help us understand ancient environments?

How were you able to predict the kind of environment in which the fossilized organism lived?

_____

_____

_____

_____

_____

What might a scientist learn about an organism by studying fossils?

_____

_____

_____

_____

_____

_____

_____

Activity 11

Observe Like a Scientist

No Bones About It!

Complete the interactive. **Choose** your organism, your material, and whether the environment includes water. As you wait for the Fossilizer 5000 to work, **predict** what will happen in the chart (fossil, partial fossil, or no fossil). After your fossil is complete, **write** whether your predictions were correct.

Quick Code: us3519s

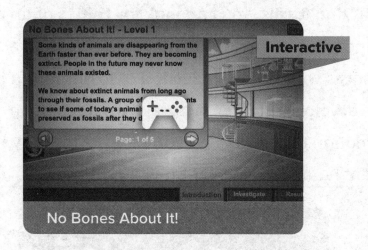

No Bones About It!

Organism + Material	Prediction	Was It Correct?

SEP Planning and Carrying Out Investigations

Concept 3.1: Understanding Fossils | 35

Activity 12
Analyze Like a Scientist

Quick Code: us3520s

Learning from Fossils

Read the text and **watch** the video about discovering and interpreting fossils. Then, **answer** the question.

Learning from Fossils

When analyzing fossils, you can look for **traits** of an organism to determine what its environment would have looked like. Paleontologists will study a collection of fossils found in one location to develop a clearer picture of an ancient environment. This information also allows paleontologists to study how an environment has changed over time.

Discovering Fossils

SEP Analyzing and Interpreting Data

Fossils of shells and marine animals found in the same location suggest that the ancient environment was aquatic. If you found fossils of flowers and **tropical** plants in an **Arctic** environment, can you describe how the environment changed over time? **Draw** what the environment looked like during the time period the fossil was alive. **Draw** what the environment looks like now.

Concept 3.1: Understanding Fossils

3.1 | Learn How can fossils help us understand ancient environments?

Activity 13

Evaluate Like a Scientist

Changing Environment

Quick Code: us3521s

The diagram shows layers of rock and other sediment. Some layers contain fossils.

Sediment	Fossils
(top layer)	sand, no fossils
	bear tracks
	frog skeletons, fresh-water snails
	shark teeth
(bottom layer)	dinosaur bones, fern-like leaves

SEP Analyzing and Interpreting Data

Circle the statement that best describes what occurred at this location, based on fossil evidence.

A. The location was dry land that became covered by a sea. The sea shrank, leaving a lake. Then, the lake dried up. As the climate became drier, the land became a desert.

B. The location was a desert for many years. The climate became wetter and wetter. First, a lake formed; then, a sea formed. Finally, the climate became drier again and the sea turned back into a lake.

C. The location has always contained water. First, a river flowed through the land. The river eroded away soil and eventually formed a lake. Then, a nearby sea flooded, joining with the lake to form an ocean.

3.1 | Learn How can fossils help us understand ancient environments?

How Can You Understand Fossil Information?

Activity 14
Observe Like a Scientist

Quick Code: us3522s

Discovering Fossils

Carefully **examine** the image of a clown fish.

Clown Fish

Complete the Change Over Time graphic organizer by describing how the fish has changed over time and the causes of those changes. Be sure to include evidence from your station activities to explain how the fish's body changed, what happened to the sediment around the fish, and how long the changes took.

SEP Analyzing and Interpreting Data
CCC Scale, Proportion, and Quantity

Topic: _____

Before:	After:

Reasons for Change:

Watch the video Discovering Fossils. After watching the video, **add** any more evidence that you can to your graphic organizer.

Discovering Fossils

Concept 3.1: Understanding Fossils | 41

3.1 | Learn How can fossils help us understand ancient environments?

Activity 15

Evaluate Like a Scientist

Quick Code: us3523s

Describe Fossils

Describe Me!

Look at the images of fossils. **Analyze** each image to determine which description matches each image. **Write** the letter of the matching description next to the image.

Descriptions:

- A. I had a long tail and lived in the ocean.
- B. I lived in the ocean and ate small fish.
- C. I had a long abdomen and lived on land and in the air.

SEP Analyzing and Interpreting Data

Fossil Images:

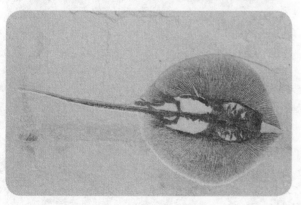

Concept 3.1: Understanding Fossils | 43

3.1 | Learn — How can fossils help us understand ancient environments?

Also Describe Me!

Extinct organisms share many of the same traits as species that are currently found on Earth. **Look** at the images and **write** the letter of the matching description next to each image.

Descriptions:

A. I have a long tail and live in the ocean.

B. I live in the ocean and eat small fish.

C. I have a long abdomen and live on land and in the air.

Images of Organisms:

Concept 3.1: Understanding Fossils | 45

3.1 | Share How can fossils help us understand ancient environments?

Activity 16

Record Evidence Like a Scientist

Quick Code: us3524s

Fossils

Now that you have learned about how fossils form and what we can learn from them, look again at the image of the fossil. You first saw this in Wonder.

Let's Investigate Fossils

Talk Together

How can you describe the fish fossil now? How is your explanation different from before?

SEP Constructing Explanations and Designing Solutions

Look again at the Can You Explain? question. You first read this question at the beginning of the lesson.

> **Can You Explain?**
>
> How can fossils help us understand ancient environments?

Now, you will use your new ideas about interactions in the environment to answer a question.

1. **Choose** a question. You can use the Can You Explain? question or one of your own. You can also use one of the questions that you wrote at the beginning of the lesson.

2. Then, **use** the information on the next page to help you answer the question.

Concept 3.1: Understanding Fossils | 47

3.1 | Share — How can fossils help us understand ancient environments?

Based on everything you have learned, **write** a scientific explanation for the Can You Explain? question or the question you chose. Be sure to **include** the three elements of a scientific explanation: a scientific claim, evidence to support the claim, and reasoning that connects the evidence to the claim.

STEM in Action

Quick Code: us3525s

Activity 17

Analyze Like a Scientist

Digging for Clues to the Past to Learn About the Present

Read the text about the work that paleontologists do. Then, **complete** the activity.

Digging for Clues

You may enjoy digging in the dirt and discovering interesting rocks, different colored soil, and maybe even worms! Paleontologists are scientists who dig for fossils in rocks, sand, soil, and dirt. They dig for fossils to uncover the secrets of the past. Fossils tell a story about what may have happened to cause a **species** to become extinct. It is the paleontologist's job to uncover the story.

As you have learned, studying fossils is important work. It helps us understand the past. It also helps us understand what the land may have looked like millions of years ago. Fossils can also reveal the type of species that lived during that time.

SEP Obtaining, Evaluating, and Communicating Information

Concept 3.1: Understanding Fossils

Digging for Clues *cont'd*

Paleontologists at Work

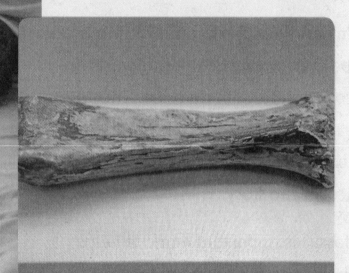
Fossilized Leg Bone

Paleontologist Gary Takeuchi digs for fossils in Red Rock Canyon State Park, in the Mojave Desert. He uncovers a 9-million-year-old lower leg bone. He shows us how he preserves it so he can study it further.

This discovery is important because the fossil of the leg bone can tell him a lot about the environment 9 million years ago. For instance, the park once had lakes, running water, and large animals. The fossil may be able to tell him why and how the landscape changed. It may also help him understand why the species died off.

Dinosaur Footprints

Look at each set of footprints. How are the dinosaur prints different from one another? What does each set of prints reveal about the characteristics of each dinosaur? **Explain** how you came to this conclusion.

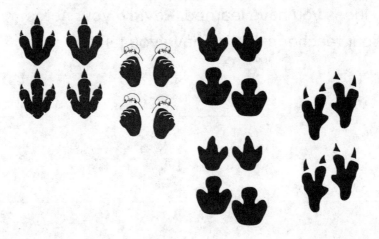

3.1 | Share — How can fossils help us understand ancient environments?

Activity 18

Evaluate Like a Scientist

Review: Understanding Fossils

Quick Code: us3526s

Think about what you have read and seen in this lesson. **Write** some key ideas you have learned. **Review** your notes with a partner. Your teacher may also have you take a practice test.

_____

_____

_____

_____

_____

_____

_____

_____

_____

_____

SEP Analyzing and Interpreting Data

Talk Together

Think about what you saw in Get Started. Use your new ideas to discuss fossils and how they can be used to provide evidence of how animals are affected by environmental changes.

CONCEPT 3.2

Interactions in the Environment

Student Objectives

By the end of this lesson:

- [ ] I can explain how an organism's environment can influence its traits and behaviors.
- [ ] I can use data to explain patterns in the distribution of organisms.
- [ ] I can investigate and describe how modifying an environment affects an organism that lives there.
- [ ] I can analyze and interpret data to explain the relationship between learning and behavior.
- [ ] I can describe patterns that predict how well different organisms can survive in particular habitats.
- [ ] I can explain how a habitat can affect organisms that live there over many generations.

Key Vocabulary

- [ ] air
- [ ] behavior
- [ ] camouflage
- [ ] coral reef
- [ ] desert
- [ ] energy
- [ ] factors
- [ ] grassland
- [ ] habitat
- [ ] instinct
- [ ] interact
- [ ] nutrient
- [ ] predator
- [ ] reproduce
- [ ] survive
- [ ] temperature
- [ ] water

Quick Code: us3528s

Activity 1
Can You Explain?

How does the environment affect where plants and animals survive?

_____
_____
_____
_____
_____
_____
_____
_____
_____
_____

Quick Code: us3529s

3.2 Wonder — How does the environment affect where plants and animals survive?

Activity 2
Ask Questions Like a Scientist

The Kangaroo Rat

Look at the image. Then, **write** your answers to the questions.

Quick Code: us3530s

Let's Investigate the Kangaroo Rat

SEP Asking Questions and Defining Problems
CCC Structure and Function

Look at the kangaroo rat's face. **List** the parts of the face that you can identify. What do you think the kangaroo rat uses each of these parts for?

_____
_____
_____
_____
_____
_____
_____
_____

Look at the kangaroo rat's long back legs. What do you think these are used for?

_____
_____
_____
_____
_____
_____
_____
_____

Concept 3.2: Interactions in the Environment | 59

3.2 | Wonder
How does the environment affect where plants and animals survive?

Now, look at the entire picture. What else do you notice about the animal and the area in which it lives?

_____
_____
_____
_____
_____
_____

Write down some other questions that you have about the kangaroo rat and its environment.

_____
_____
_____
_____
_____
_____
_____
_____
_____

Activity 3

Observe Like a Scientist

Surviving in Habitats

Think about where you live and the things in your environment. Then, **draw** a picture of your home and the environment around it.

Quick Code: us3531s

Now, **share** one thing about your drawing with a neighbor. **Talk** about why you included it.

Watch the videos and **answer** the questions.

Habitats

Home for Living Things

Concept 3.2: Interactions in the Environment

3.2 | Wonder
How does the environment affect where plants and animals survive?

Work with your classmates to create a definition for the word *habitat*. Be sure to include information about the different things that a habitat provides to the organisms that live there.

_____

_____

_____

_____

_____

Look again at the drawing you made. Go back and **add** some things to the drawing that meet your needs. On the drawing, **label** objects in the environment that meet your needs and **describe** how they meet your needs.

Look at the image and **identify** how the needs of the living things are being met by their habitat. Then, **answer** the questions.

Neighborhood

How are the needs of living things being met in this neighborhood?

_____
_____
_____
_____
_____
_____

How can living and nonliving things interact in the neighborhood?

_____
_____
_____
_____
_____
_____

3.2 | Wonder How does the environment affect where plants and animals survive?

Activity 4
Investigate Like a Scientist

Quick Code: us3532s

Hands-On Investigation: Living or Nonliving?

What is the difference between living and nonliving things?

In this activity, you will **investigate** this question by going outside to make observations of both living and nonliving things.

Make a Prediction

What are some differences that you think you will be able to observe between living and nonliving things in a habitat?

_____
_____
_____
_____
_____
_____
_____
_____

What materials do you need? (per group)

- Pencils
- Access to a natural area

What Will You Do?

1. Before you make any observations, you need to design a recording sheet to record your observations. Make sure your recording sheet includes space to record your observations about both living and nonliving things and how they interact with each other.

2. Go outside and collect data on living and nonliving things in a nearby habitat.

3. Record your observations on your recording sheet.

4. Work with your group to analyze, discuss, and display your findings.

5. Share your observations with the class.

3.2 | Wonder — How does the environment affect where plants and animals survive?

Think About the Activity

Did your original plan for your recording sheet work? What kinds of problems did you run into while making your recording sheet? How did your solution change the recording sheet?

_____
_____
_____
_____
_____
_____

How did you compare living things and nonliving things interacting with each other in their environments?

_____
_____
_____
_____
_____
_____

Explain the differences between living and nonliving things.

3.2 | Wonder How does the environment affect where plants and animals survive?

Activity 5
Evaluate Like a Scientist

Quick Code: us3533s

What Do You Already Know About Interactions in the Environment?

Plant Habitats

Look at the list of items that are found in a habitat, and then **match** each with the need it meets for flowering plants. **Write** the name of the item under the need that it meets. Then, **answer** the questions.

| rain | sunlight | soil | bees | space |

Needs	Items
Room to Grow	
Help with Reproduction	
Water	
Energy	
Minerals	

68

What experiences have you had working with plant habitats?

___

Think of some plants that live in your local habitat. What elements in the habitat do they depend on to survive? **Write down** your ideas and then share with your neighbor and discuss.

___

Concept 3.2: Interactions in the Environment

3.2 | Wonder
How does the environment affect where plants and animals survive?

Animal Habitats

Look at the photograph of a squirrel in its habitat.

Squirrel

List four things that the squirrel gets from its habitat.

_____

_____

_____

_____

Now, **think** about an animal that lives in your local environment. **List** at least four things that it gets from its habitat.

_____

_____

_____

_____

3.2 | Wonder
How does the environment affect where plants and animals survive?

Water Habitats

Which of these does a fish need to survive? **Circle** all correct answers. Then, answer the question.

Water

Food

Oxygen

Sunlight

Soil

Desert Habitats

Read the question. **Think** about how plants might become adapted to live in different types of environments. Then, **share** your ideas with a neighbor and **discuss** your answers. When you think you know the answer, **circle** it.

What is the main way in which desert plants are different from plants that live in other environments? **Circle** the correct answer.

A. They need more water.

B. They need less water.

C. They need less sunlight.

D. They need more sunlight.

Concept 3.2: Interactions in the Environment

3.2 | Learn How does the environment affect where plants and animals survive?

How Can You Describe a Habitat?

Activity 6
Observe Like a Scientist

Quick Code: us3534s

Cause and Effect

Look at the image and **think** about cause and effect. Then, **complete** the first row of the Cause/Effect Chart.

A Bad Sunburn

CCC Cause and Effect

Topic: _____

Cause → Effect

Cause → Effect

Cause → Effect

Concept 3.2: Interactions in the Environment

3.2 | Learn — How does the environment affect where plants and animals survive?

Now, **look** at the two pictures. **Think** about the cause-and-effect relationship that each picture shows. Record your observations using the Cause/Effect Chart on the previous page.

A girl wearing a wool hat and scarf puts on gloves.

A very large dog sits in the grass.

Activity 7

Observe Like a Scientist

Habitat Characteristics

Quick Code: us3535s

Complete the interactive Habitat Characteristics. **Answer** the questions while you complete the interactive.

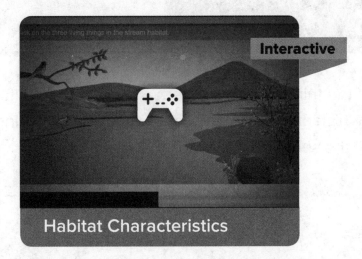

Habitat Characteristics

First, **write** three living things that you can find in a stream habitat.

_____

_____

_____

SEP Obtaining, Evaluating, and Communicating Information
CCC Systems and System Models

Concept 3.2: Interactions in the Environment

3.2 | Learn   How does the environment affect where plants and animals survive?

Next, **write** the habitat that each animal would be likely to live in:

Polar Bear	
Earthworm	
Octopus	

Make a list of the things you observed in the stream habitat. **Think** about which are living things and which are nonliving things. **List** all the subcomponents of the habitat system.

_____

_____

_____

_____

_____

How is this stream habitat system similar to or different from the habitat system you live in?

Concept 3.2: Interactions in the Environment

Activity 8
Analyze Like a Scientist

Quick Code: us3536s

Parts of a Habitat

Read the text and **answer** the questions.

Parts of a Habitat

A **habitat** is a part of the environment that provides what living things need to **survive**. Look at the squirrel image. Picture the environment where this animal lives.

Squirrel Eating an Acorn

SEP Obtaining, Evaluating, and Communicating Information

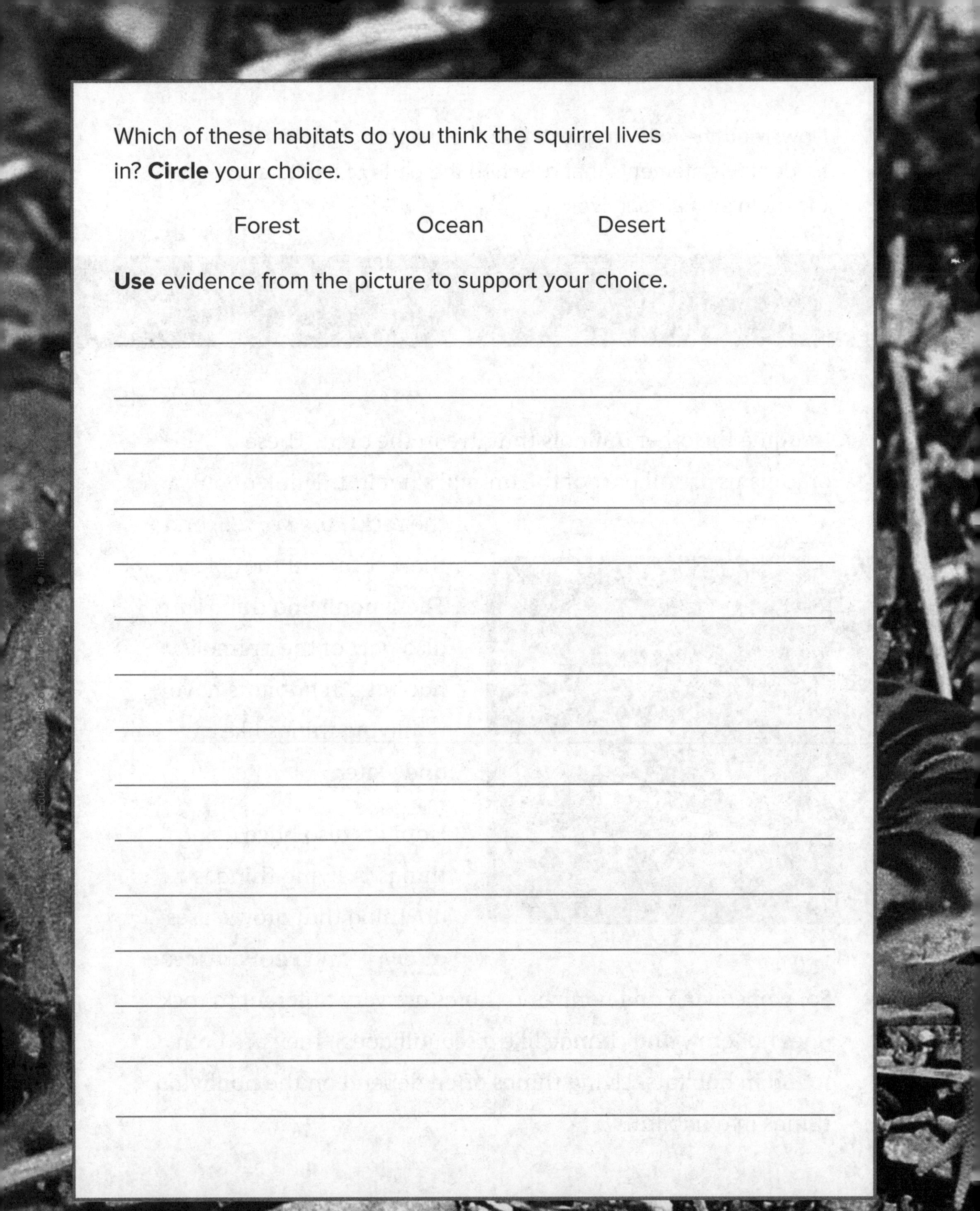

Which of these habitats do you think the squirrel lives in? **Circle** your choice.

Forest   Ocean   Desert

Use evidence from the picture to support your choice.

_____
_____
_____
_____
_____
_____
_____
_____
_____
_____
_____
_____
_____

Concept 3.2: Interactions in the Environment

Now, **read** the following text. **Use** the graphic organizer to identify statements that describe the parts of habitats that help animals survive.

Imagine the other animals that live in the area. These organisms are all part of the animal's habitat. Think about the rocks, the **water**, and the weather in that place. These nonliving things are also part of the animal's habitat. All habitats have nonliving things like **air** and water.

Habitat

Habitats also have living things. A living thing is anything that grows, uses **energy**, and **reproduces**. So, while living and nonliving things are very different (a rock does not grow and change like a seagull does), they are both found in habitats. Living things often depend on the nonliving things in a habitat.

Definition	**Examples** (drawn or written)
• Personal:	
• Dictionary:	

Habitats

Sentences	**Related:**
• Teacher/Book:	
• Personal:	

Outside of School (Who would use the word? How would they use it?):

When you have completed your graphic organizer, **share** your ideas with a neighbor. **Notice** similarities and differences between your observations.

Concept 3.2: Interactions in the Environment

Now, **answer** the question: Could the squirrel live in the desert or the ocean? Use evidence from the reading to explain your answer.

_____

_____

_____

_____

_____

_____

_____

_____

_____

_____

_____

_____

Activity 9
Think Like a Scientist

Neighborhood Habitat

Quick Code: us3537s

In this investigation, you will **explore** the habitats in your area to look for organisms and the ways that their needs are met in their habitats.

What materials do you need? (per group)

- Hand lens
- Colored pencils
- Camera

What Will You Do?

1. You are going to take a walk outside to observe how plants and animals interact with their habitat.

2. Write down a question you will try to answer from your observations.

_____

_____

SEP Planning and Carrying Out Investigations

Concept 3.2: Interactions in the Environment | 85

3.2 | Learn How does the environment affect where plants and animals survive?

3. Collect your observations. Make sketches of what you see.

4. Share your observations with other students. Work in a group to organize your data into the ways the living thing in the habitat met its needs (food, water, shelter).

5. Categorize your observations of the habitat into sources of food, water, and shelter for the organisms that live there.

6. Explain how humans can change the habitat in a positive way and a negative way.

Think About the Activity

What is one way that humans could change the habitat in a positive way? How would this affect the habitat in the future?

_____
_____
_____
_____
_____
_____

What is one way that humans could change the habitat in a negative way? How would this affect the habitat in the future?

_____
_____
_____
_____
_____
_____

Concept 3.2: Interactions in the Environment

3.2 | Learn — How does the environment affect where plants and animals survive?

How Can the Environment Affect Living Things?

Activity 10

Think Like a Scientist

Plant Adaptations

Quick Code: us3538s

In this activity, you will **observe** two different types of plants and identify the natural habitat of each plant.

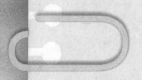

What materials do you need? (per group)

- Live cactus
- Live *Elodea*
- Containers to hold the plants
- Index cards

What Will You Do?

1. Closely observe a cactus and an *Elodea*.
2. Identify the traits of each plant that allow it to survive in its particular environment.

SEP Planning and Carrying Out Investigations

Think About the Activity

Plant Observations		
Plant Name	**Sketch of Plant**	**Features of the Plant that Help It Survive in Its Environment**

Concept 3.2: Interactions in the Environment

3.2 | Learn How does the environment affect where plants and animals survive?

What do you think would happen if the plants switched habitats? Could they survive? Why or why not?

_____
_____
_____
_____

What are some of the characteristics of the cactus and *Elodea* that allow them to adapt to their particular environments?

_____
_____
_____
_____

Activity 11

Analyze Like a Scientist

Quick Code: us3539s

Meeting Needs

Watch the video. Then, **read** the text and **look** at the images. **Discuss** and **answer** the questions as you watch and read.

Need for Food and Water

Meeting Needs

All plants need light, water, and some other **nutrients** to grow. Plants grow best when these are available in the amounts they need.

Concept 3.2: Interactions in the Environment | 91

Meeting Needs cont'd

Redwood Forest

Arizona Squirrel

Idaho Squirrel

Lack of one of these needs can change the way a plant grows. For example, trees that grow in the shade of big trees are often stunted and thin because they do not get enough sunlight. Only when the big tree dies will they get enough light to grow properly.

An animal can be affected by its environment, too. Squirrels need to live in a habitat where they can easily find nuts for food. A squirrel that can find food easily will grow larger than a squirrel in a habitat with less access to a food source. Even though the animals are the same, their size may vary based on where they live.

How do the plants found on the floor of the redwood forest obtain the energy they need?

_____
_____
_____
_____
_____
_____

What might prevent the plants on the floor of the redwood forest from obtaining the energy they need?

_____
_____
_____
_____
_____
_____
_____

How do the redwood trees get the water they need to survive?

_____
_____
_____
_____
_____
_____

What would happen if we planted a redwood tree in the desert? How do you know?

_____
_____
_____
_____
_____
_____
_____
_____

How does a squirrel living in Arizona or Idaho obtain the energy it needs?

_____
_____
_____
_____
_____
_____

What might prevent a squirrel living in Arizona or Idaho from obtaining the energy it needs?

_____
_____
_____
_____
_____
_____

3.2 | Learn How does the environment affect where plants and animals survive?

Activity 12
Observe Like a Scientist

Quick Code: us3540s

Learning

Watch the video to learn about how animals learn behaviors that help them survive.

Survival Behavior

Next, **complete** the interactive. As you complete the interactive, **answer** the questions.

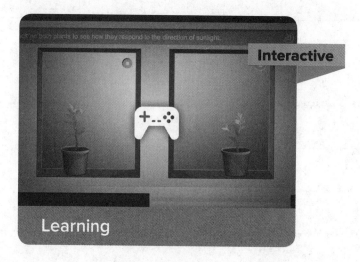

Learning

What does the word *behavior* mean?

_____

_____

_____

Is crying a learned behavior? Is reading a learned behavior? **Explain** your answers.

_____

_____

_____

_____

Describe how plants respond to sunlight.

_____

_____

_____

_____

Concept 3.2: Interactions in the Environment

How Can an Animal Learn Behavior from Its Parents and the Environment?

Activity 13
Analyze Like a Scientist

Quick Code: us3541s

Animal Behavior

What you have read and seen about animals, behaviors, and habitats so far? **Record** some thoughts about what you have learned in the chart. As you read, **add** information to your chart.

AEIOU	New Learning
Adjective	
Emotion	
Interesting	
Oh!	
Um	

98

Read the text about inherited and learned behaviors. As you read, **look** for evidence of learned and inherited behaviors. **Write** down your evidence in the graphic organizer that follows.

Animal Behavior

Animals have natural abilities and **behaviors**. Behaviors that come naturally, without learning, are inherited from the animal's parents. Inherited behaviors are behaviors an animal is born with. A newborn puppy knows to crawl to its mother for milk.

Inherited behaviors are sometimes called **instincts**. These are the behaviors that often help an animal survive. A spider spinning a web is another example of an inherited behavior. A spider does not learn how to spin a web from its parents. Spinning a web is an instinct.

Spider's Web

Animal Behavior *cont'd*

Dog with a Ball

Wild Dogs

Learning happens when an organism gains information or experience that changes its behavior. A dog is not born knowing how to fetch a ball. To train a dog to fetch, a person may give the dog a treat each time it successfully picks up and returns the ball. The dog learns that fetching a ball will lead to a reward, and fetching becomes part of the dog's behavior.

An animal's behavior can be influenced by many **factors**. One way is through learned behaviors from parents. Another factor that influences behavior is the environment. As an animal **interacts** with its environment, learning can occur. A dog that walks through the same neighborhood everyday with its owner soon learns which neighbors hand out treats, and the dog will show signs of excitement every time they walk by that house. In the wild, animals interact with the environment to learn which foods are edible, or where water regularly pools after a rainstorm. These interactions with the environment change the animal's behavior.

Keep in mind that both inheritance and environment influence animals' behaviors, but animals also have limits on what they can learn. A spider cannot follow commands, and a dog could not learn how to read. But a dolphin can learn how to jump through hoops, because it is smart enough and because it has the strength to do it.

Changes in the environment can also produce changes in behavior. For example, squirrels store their food to prepare for winter, when food becomes scarce. A gentle dog may become hostile when other animals get into the house. Behavior is the result of inherited traits, learning, and the environment.

Topic: _____

Inherited	Learned

Activity 14
Evaluate Like a Scientist

Quick Code: us3542s

Learned vs. Inherited

A scientist has made these observations in animals. **Analyze** these data to determine if the behavior is learned or inherited. **Write** an **L** by each behavior that is learned and an **I** by each behavior that is inherited.

Chimpanzees sharpen sticks and use them as tools for hunting.

Mother dogs nurse their babies immediately after they are born.

A bird avoids eating a certain kind of insect after it gets sick from eating that insect.

A dolphin in captivity jumps through hoops and is rewarded with a fish.

Bears cannot find enough food to survive in winter, so they hibernate in caves.

SEP Analyzing and Interpreting Data

3.2 | Learn How does the environment affect where plants and animals survive?

Activity 15
Observe Like a Scientist

Types of Habitats

Quick Code: us3543s

Watch the videos. **Think** about the parts of habitats that you identified earlier. As you watch the videos, **look** for evidence of each part.

Water Habitats

Desert Habitats

- **SEP** Obtaining, Evaluating, and Communicating Information
- **CCC** Systems and System Models
- **CCC** Patterns

Grasslands

Now, **list** the parts of habitats that you identified. Include specific examples of each part that you observed in the videos.

3.2 | Learn How does the environment affect where plants and animals survive?

What patterns can you identify across all three videos?

_____
_____
_____
_____
_____
_____

Do you think an animal that lives in the **grassland** could also live in the desert? **Explain** why or why not.

_____
_____
_____
_____
_____
_____
_____
_____
_____

How Do Habitats Match the Needs of Many Living Things?

Activity 16

Observe Like a Scientist

Quick Code: us3544s

Foxes

These pictures show three types of foxes: a fennec fox, an Arctic fox, and a red fox. **Look** carefully at each image, and **think** of some questions that you have about them.

Fennec Fox

Arctic Fox

Concept 3.2: Interactions in the Environment

3.2 | Learn — How does the environment affect where plants and animals survive?

Red Fox

Write a question you have after closely analyzing these three images.

_____
_____
_____
_____
_____
_____

Share your question with a partner, and work together to **find** two pieces of evidence in the images that can help answer your question. Be sure to **help** your partner answer their question, too!

Write the two pieces of evidence you found to answer your question.

_____
_____
_____
_____
_____

What are some ways that each fox is suited to live in its environment?

_____
_____
_____
_____
_____
_____
_____

3.2 | Learn   How does the environment affect where plants and animals survive?

What would happen if the Arctic fox had to live in the desert?

_____

_____

_____

_____

_____

_____

_____

_____

_____

_____

_____

_____

SEP Asking Questions and Defining Problems
CCC Patterns

Activity 17
Analyze Like a Scientist

Organisms in Habitats

Quick Code: us3545s

Read the text and then **answer** the questions.

Organisms in Habitats

Habitats can be very different from one another. As a result, animals and plants in these habitats can be very different. For example, some **desert** plants have spiny prickles instead of flat leaves. A desert habitat is dry and has little water, so the spiny leaves help reduce the amount of water plants lose to evaporation. Polar bears have thick layers of fat that help keep them warm in the cold environment.

Polar Bear

SEP Constructing Explanations and Designing Solutions

Concept 3.2: Interactions in the Environment | 111

Organisms in Habitats *cont'd*

Animals need a way to protect themselves from **predators**. A habitat can provide protection. A habitat might have hiding places for animals. Plants can provide a hiding spot for small animals so the animals can avoid a predator. The **coral reef** provides many hiding places for fish. If a coral reef was destroyed, what would happen to the fish that need the reef for protection?

Fish in Coral

An animal might also have fur or skin color that can provide **camouflage** in its habitat. If an animal is not able to find good protection against predators, that species will not survive as well. An animal must have all of its needs met within a habitat to survive.

What would happen if a bison was placed in the Arctic Sea and a polar bear was placed in a grassland?

_____

_____

_____

_____

How does the Arctic Sea meet the needs of the polar bear?

_____

_____

_____

_____

Would the polar bear be able to eat in a grassland?

_____

_____

_____

_____

Concept 3.2: Interactions in the Environment

How would the **temperature** range found across the grassland and Arctic affect the polar bear and bison?

___

What problems would the bison have living in the Arctic?

___

Activity 18
Analyze Like a Scientist

Quick Code: us3546s

Kangaroo Rat

Read the text about the kangaroo rat. As you read, **underline** any part of a sentence that describes a part of the kangaroo rat's body. **Circle** any part of a sentence that describes the behavior of the kangaroo rat.

Kangaroo Rat

Kangaroo rats are small mammals that live in the desert. They have large heads with big eyes. They have long tails and large hind legs. They are sandy brown on top and white on the bottom.

Kangaroo Rat

SEP Constructing Explanations and Designing Solutions
CCC Cause and Effect

Concept 3.2: Interactions in the Environment | 115

Kangaroo Rat *cont'd*

Kangaroo rats' unique traits are adapted to life in the desert. Water is rare in the desert, so many of the kangaroo rat's adaptations involve preserving water. In fact, they can go their entire lives without drinking liquid water because they can get enough water from the seeds they eat. They do not pant or sweat to cool themselves. If they did, they would lose water. Instead, they sleep underground during the day. When it is cool at night, they go out and search for food.

Kangaroo rats have many adaptations to escape potential predators. They have excellent hearing, so they can hear predators approaching before they get too close. Kangaroo rats can also jump up to 9 feet high to escape predators. Additionally, their sandy brown coloring helps them blend into the desert sands.

Kangaroo rats have pouches. However, these pouches are not for carrying babies. The pouches are in their cheeks, and the kangaroo rats use them to carry seeds back to their burrows.

Now, go back and **look** at what you circled and underlined. **Identify** how the specific characteristics and behaviors of the kangaroo rat meet its needs. **Record** your observations using the chart.

Topic: _____

Characteristic	Meets Its Need

Concept 3.2: Interactions in the Environment

Using evidence from the text, **explain** why the kangaroo rat is best suited to survive in a desert habitat.

Activity 19
Think Like a Scientist

Habitats Are All Around

Quick Code: us3547s

In this activity, you will choose a habitat to research. For example, you could choose the desert, ocean, bay, mountains, or forest. You will research the plants and animals that live in that habitat.

What materials do you need? (per group)

- Pencils
- Research materials such as:
  - Internet access
  - Library access
  - Relevant books and magazines
- Writing paper or science notebook for recording ideas

What Will You Do?

1. Research the characteristics of your chosen habitat and the plants and animals that live there.

2. Choose a specific organism that lives in the habitat, and research the ways in which the habitat meets the organism's specific needs.

SEP Constructing Explanations and Designing Solutions
CCC Cause and Effect

3.2 | Learn — How does the environment affect where plants and animals survive?

Think About the Activity

Describe the characteristics of the habitat you selected. Compare this habitat to another habitat. You can include images, graphs, charts, tables, or diagrams to support your answer.

_____
_____
_____
_____
_____
_____
_____
_____
_____
_____

What animals live in the habitat? What roles do they play in the ecosystem?

_____
_____
_____

What plants live in the habitat? What roles do they play in the ecosystem?

Select one of the animals or plants that live in the habitat you selected. **Explain** how the habitat meets that organism's needs.

Concept 3.2: Interactions in the Environment

3.2 | Learn — How does the environment affect where plants and animals survive?

Activity 20
Evaluate Like a Scientist

Adaptations and Habitats

Quick Code: us3548s

Adaptations to Survive

Look at the image of each habitat, and **write** the letter of the trait that is best suited for that habitat next to the image.

Traits:

A. An animal has thick, white fur.

B. A plant has structures that attach firmly to rocks.

C. A plant has long roots to reach water underground.

D. An animal has a long bill to drink nectar.

Habitats:

Arctic

_____

CCC Cause and Effect

Desert

Tropical Flower

Rocks Near Ocean

Concept 3.2: Interactions in the Environment | 123

3.2 | Learn — How does the environment affect where plants and animals survive?

Where Is My Habitat?

The map below is a topographical map of California that shows mountain ranges, deserts, coastlines, and outlying islands.

Look at the map and read the descriptions of the plants and animals. Then, **write** the number of the habitat where the animal would live next to the description.

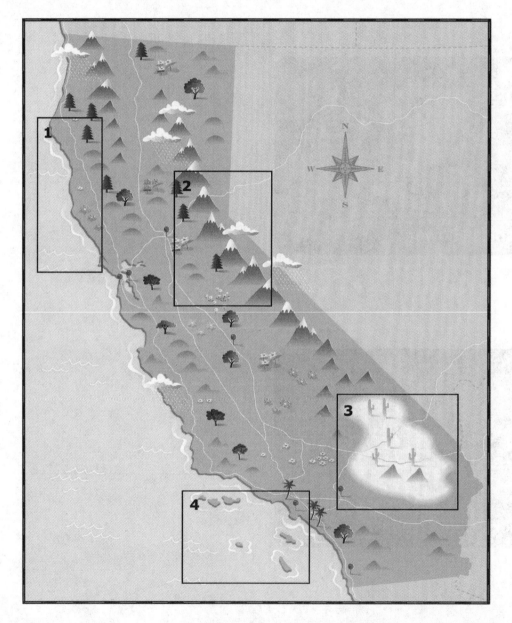

This plant has long, deep roots and a waxy coating on its surface. It needs water, sunlight, air, and soil.

_____

This animal is surefooted and eats grasses and woody plants. Its coloring changes during the year, especially when it loses its heavy winter coat. It needs water, air, and food.

_____

This tree is exceptionally tall and needs large amounts of water and fog to grow. It also needs sunlight, air, and soil.

_____

This animal is able to swim very quickly to catch its prey. It needs food, air, and water to survive.

_____

Explain Your Reasoning

Choose one of the organisms from the previous activity. **Explain** why you made the connection to the habitat you picked.

_____

_____

_____

_____

_____

_____

Concept 3.2: Interactions in the Environment

3.2 | Share How does the environment affect where plants and animals survive?

Activity 21

Record Evidence Like a Scientist

Quick Code: us3549s

The Kangaroo Rat

Now that you have learned about interactions between living things and their environment, look again at the image of the kangaroo rat. You first saw this in Wonder.

Let's Investigate the Kangaroo Rat

Talk Together

How can you describe the kangaroo rat now? How is your explanation different from before?

SEP Constructing Explanations and Designing Solutions

Look at the Can You Explain? question. You first read this question at the beginning of the lesson.

> **Can You Explain?**
>
> How does the environment affect where plants and animals survive?

Now, you will use your new ideas about interactions in the environment to answer a question.

1. **Choose** a question. You can use the Can You Explain? question or one of your own. You can also use one of the questions that you wrote at the beginning of the lesson.

My Question

2. Then, **use** the graphic organizers to help you answer the question.

Concept 3.2: Interactions in the Environment

3.2 | Share — How does the environment affect where plants and animals survive?

To plan your scientific explanation, first **write** your claim. Your claim is a one-sentence answer to the question you investigated. It answers: *What can you conclude?* It should not start with *yes* or *no*.

My claim:

Data 1

Data 2

Finally, **explain** your reasoning. Reasoning ties together the claim and the evidence. Reasoning shows how or why the data count as evidence to support the claim.

Topic: _____

Evidence	How It Supports the Claim

Concept 3.2: Interactions in the Environment

3.2 | Share — How does the environment affect where plants and animals survive?

Now, **write** your scientific explanation.

The environment affects where plants and animals survive because . . .

_____
_____
_____
_____
_____
_____
_____
_____
_____
_____
_____
_____
_____
_____
_____

STEM in Action

Quick Code: us3550s

 Activity 22
Analyze Like a Scientist

Careers and Habitat Characteristics

Read the text about habitat restoration specialists and **watch** the videos. Then, **answer** the questions.

Careers and Habitat Characteristics

Humans change many habitats when we add our buildings and roads. Habitat restoration specialists play a very important role. They help bring back important plant and animal habitats.

Habitat Plan

SEP Asking Questions and Defining Problems

Concept 3.2: Interactions in the Environment

Careers and Habitat Characteristics *cont'd*

This work is especially important for endangered species. The loss of habitat is one of the leading causes of extinction. Restoring habitat can help solve problems for species in trouble.

Molly and Dave from the show *Backyard Habitat* help viewers understand the important role of habitat restoration specialists. They also help people restore and create important animal habitats close to their homes. Watch how Molly and Dave plan to restore a section of the yard to support a habitat for the anole lizard.

A New Landscape

Your Backyard Habitat

What plant or animal would you like to make a habitat for around your home?

_____

_____

_____

_____

Explain how you could make a habitat that would fulfill all of its needs.

_____

_____

_____

_____

_____

_____

_____

3.2 | Share   How does the environment affect where plants and animals survive?

Activity 23

Evaluate Like a Scientist

Quick Code: us3551s

Review:
Interactions in the Environment

Think about what you have read and seen in this lesson. **Write** some key ideas you have learned. **Review** your notes with a partner. Your teacher may also have you take a practice test.

_____
_____
_____
_____
_____
_____
_____
_____
_____
_____

Talk Together

Think about what you saw in Get Started. Use your new ideas to discuss how animals are affected by environmental changes.

CONCEPT
3.3

Environmental Changes

Student Objectives

By the end of this lesson:

- [ ] I can ask questions about how human activities affect the environment.
- [ ] I can explain how the rate of change in an environment depends on certain factors.
- [ ] I can use logical reasoning to predict the effects of changes in environments on the organisms that live there.
- [ ] I can develop a model that explains how living and nonliving parts of the environment affect each other.
- [ ] I can argue from evidence that changes in land ecosystems affect water ecosystems on small and large scales.

Key Vocabulary

- [ ] adjust
- [ ] community
- [ ] ecosystem
- [ ] impact
- [ ] natural
- [ ] pollution
- [ ] recycle
- [ ] survival

Quick Code: us3553s

Activity 1
Can You Explain?

How does the environment change over time?

_____
_____
_____
_____
_____
_____
_____
_____
_____

Quick Code:
us3554s

3.3 | Wonder — How does the environment change over time?

Activity 2
Ask Questions Like a Scientist

The Salt Harvest Mouse

Quick Code: us3555s

Look at the image.

Let's Investigate the Salt Harvest Mouse

SEP Asking Questions and Defining Problems

Look at the image of the salt harvest mouse. Use the KWL Chart to record your observations about the salt harvest mouse. In the first column (What I Know), record any facts or inferences you can make from looking at this image. In the second column (What I Want to Know) write any questions you have about the salt harvest mouse. Throughout this lesson, you can add to the KWL Chart as you learn new information that can help you answer your questions.

Topic: _____

What I **K**now	What I **W**ant to Know	What I Have **L**earned

Concept 3.3: Environmental Changes | 141

Activity 3

Analyze Like a Scientist

Changes to Habitats

Quick Code: us3556s

Read the text about the habitat of the salt harvest mouse and **watch** the video. Then, **answer** the questions.

Changes to Habitats

People **adjust** to changes all the time. When it feels cold outside, you put on a jacket. When you start in a new class, you learn about your new teacher and new rules. You constantly adjust your behavior to deal with changes. Plants and animals also change when their environment changes.

Salt Harvest Mouse

A salt harvest mouse's habitat is a marsh. Its habitat has changed over time, making **survival**

CCC Cause and Effect

for the mouse more difficult. What changes can **impact** an animal's survival?

There can be **natural** changes over time and quicker changes created by people. How can change be harmful or helpful to an environment?

Changes in Habitats

What would be a change to the habitat that would be a positive change for the salt harvest mouse?

_____

_____

_____

_____

What other predictions can you make about how changes to the habitat could affect the salt harvest mouse?

_____

_____

_____

_____

Activity 4
Evaluate Like a Scientist

Quick Code: us3557s

What Do You Already Know About Environmental Changes?

Read the following list of human activities. Then, **draw** a line connecting each activity with the environmental change it causes.

Human Activity
Dumping pollution in a river
Paving a grassland
Planting trees
Cutting a forest
Cleaning a river

Environmental Change
Improves water quality for fish and aquatic plants
Kills grasses and removes insects' habitat
Destroys trees and removes animals' habitat
Kills fish and aquatic plants
Adds shade and new habitat for animals

Concept 3.3: Environmental Changes | 145

3.3 | Wonder — How does the environment change over time?

Forest Fires

Some ecosystems experience wildfires that change the landscape very quickly. In the space, **write** your ideas about how plants and animals in a forest could be affected by a forest fire.

_____
_____
_____
_____
_____
_____
_____
_____
_____
_____

3.3 | Learn How does the environment change over time?

How Can Environments Change?

Activity 5
Observe Like a Scientist

Long-Term Changes in Ecosystems

Quick Code: us3558s

The environment around us is constantly changing. Some changes happen quickly while others happen very slowly. In this interactive activity, you will examine evidence of climatic changes that can lead to changes in ecosystems over time. As you complete the interactive, **use** the Cause/Effect graphic organizer to record your evidence.

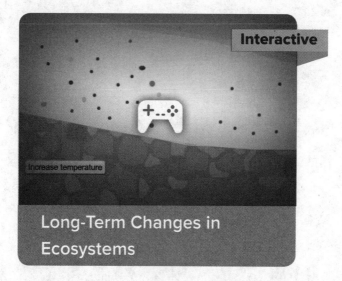

Long-Term Changes in Ecosystems

CCC Stability and Change
CCC Cause and Effect

Topic: _____

Cause	Effect

Answer the questions based on what you learned in the interactive activity:

How does an ecosystem change when a glacier melts and moves back?

_____
_____
_____
_____

Concept 3.3: Environmental Changes | **149**

3.3 | Learn How does the environment change over time?

How does a forest ecosystem grow and develop after a glacier retreats?

_____

_____

_____

_____

Activity 6
Analyze Like a Scientist

Changes to the Environment

Read the text with your class. As you read, **raise your hand** whenever you hear about a type of change (such as a flood, earthquake, or fire) that you have personally observed.

Changes to the Environment

Environments are changing constantly. Some changes happen slowly over time, while other changes can happen quickly.

Flooding Stream

Over long periods of time, many different factors can affect and change **ecosystems**. Mountains are formed and worn away. Rivers change course. Storms can cause flooding, and earthquakes can cause damage to the land.

SEP **Obtaining, Evaluating, and Communicating Information**
CCC **Stability and Change**   CCC **Cause and Effect**

Concept 3.3: Environmental Changes | 151

Changes to the Environment *cont'd*

Changes in Earth's climate cause temperatures to change, resulting in sea levels rising or falling. All of these natural changes affect both living and nonliving parts of an ecosystem.

Many human activities affect ecosystems. Building, farming, mining, and logging disturb land habitats. Building dams, changing waterways for transportation or construction, and even getting drinking water disturb water habitats.

Digging a Hole

Unlimited hunting and overfishing remove many animals from an ecosystem. Human activities can result in **pollution** and harm ecosystems. We can pollute the air or water, which can poison and kill animals and plants.

Talk Together

Get together with a partner and share one human-caused and one natural environmental change that you have identified. Determine if each change happens quickly or slowly.

Concept 3.3: Environmental Changes

3.3 | Learn How does the environment change over time?

Activity 7
Observe Like a Scientist

Quick Code: us3560s

Impacts on Earth

Work in small groups to **watch** and **discuss** the videos. As you watch each video, **identify** whether each change is natural or caused by humans. After you watch each video, **answer** the question about it that follows.

Impacting Earth

SEP Obtaining, Evaluating, and Communicating Information
CCC Cause and Effect

Floods

Is this change natural or caused by humans? How might it affect local organisms?

_____
_____
_____
_____
_____
_____

3.3 | Learn How does the environment change over time?

Pollution in the Ocean

Is this change natural or caused by humans? How might it affect local organisms?

_____
_____
_____
_____
_____
_____

Is this change natural or caused by humans? How might it affect local organisms?

_____

_____

_____

_____

_____

Concept 3.3: Environmental Changes

3.3 | Learn — How does the environment change over time?

Activity 8
Evaluate Like a Scientist

Natural or Human Made?

Quick Code: us3561s

Some of the environmental changes listed here are caused by humans, and others have natural causes. **Circle** all those that are caused by humans.

- Wildfires
- Flooding
- Building roads
- Volcanic eruptions
- Cattle grazing
- Farming
- Earthquake
- Cutting down trees
- Desert
- Drought

CCC Cause and Effect

Activity 9
Observe Like a Scientist

Population Changes

Quick Code: us3562s

Complete the interactive. As you work, **answer** the questions.

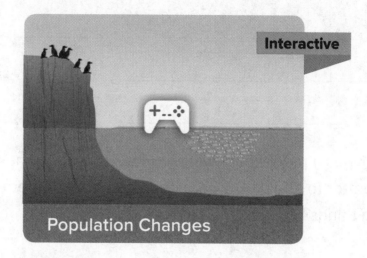

How can a change in the climate affect the population of a species?

_____
_____
_____
_____
_____

SEP Developing and Using Models
CCC Cause and Effect

Concept 3.3: Environmental Changes

3.3 | Learn   How does the environment change over time?

Why does a change in the population of one species affect the population of other species?

_____
_____
_____
_____
_____

The interactive is a model of a system. How could you make this model better to show how changes in an environment impact living things?

_____
_____
_____
_____
_____

What Happens to Living Things When an Environment Changes?

Activity 10
Observe Like a Scientist

Quick Code: us3563s

Short-Term Changes

In this activity, you will **observe** changes in a forest ecosystem caused by several types of human activity. **Complete** the interactive and **answer** the questions.

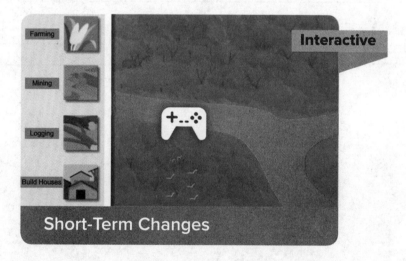

CCC Cause and Effect

Concept 3.3: Environmental Changes

3.3 | Learn How does the environment change over time?

In what ways do human activities like farming, mining, logging, and building affect animal populations in an ecosystem?

_____

_____

_____

How can humans farm, log, mine, and build houses with less damage to the ecosystem?

_____

_____

_____

Activity 11

Think Like a Scientist

A Changing Environment

Quick Code: us3564s

In this activity, you will compare models of a current environment and the same environment 100 years ago to identify ways that humans have caused changes to the environment.

What materials do you need? (per group)

- Internet access for research and/or mixed media (magazines, nonfiction texts) about the area
- Paper for drawing
- Pencils
- Colored pencils

SEP Obtaining, Evaluating, and Communicating Information
CCC Patterns

3.3 | Learn How does the environment change over time?

What Will You Do?

1. First, choose an environment that exists today.

2. Draw a picture of it. Include as much detail as you can about the living and nonliving things that are part of the environment.

3. Research what the same environment looked like 100 years ago. Based on your research, draw a model showing the environment 100 years ago.

4. Compare your models to identify ways that the environment has changed in the past 100 years.

Today

100 Years Ago

What environment did you choose? How has this environment changed in 100 years?

_____
_____
_____
_____
_____
_____

Concept 3.3: Environmental Changes

3.3 | Learn How does the environment change over time?

Think About the Activity

Compare your drawing to another student's. How are they different? What did you both include in your drawing? Why do you think your drawings differ?

_____
_____
_____
_____
_____
_____

What do you notice about the change in natural resources (living and nonliving) over the last 100 years?

_____
_____
_____
_____
_____
_____

Analyze your observations to **explain** why the living things in the past environment are different from what is there now.

_____
_____
_____
_____
_____
_____
_____

How can humans limit their impact on the natural environment in which they live?

_____
_____
_____
_____
_____
_____
_____

Concept 3.3: Environmental Changes

Activity 12
Analyze Like a Scientist

Quick Code: us3565s

Impacts of Habitat Change

Read the passage describing how changes impact habitats. As you read, **highlight** one sentence in the text and **explain** how it connects to something you learned earlier. Then, **share** your ideas with a classmate and determine if your connection statements are related to each other.

Impacts of Habitat Change

Grizzly Bear in Its Natural Habitat

All the parts of an ecosystem interact with one another. Living things in an environment can survive in a particular habitat because their needs are met well.

SEP Obtaining, Evaluating, and Communicating Information

Organisms are affected when a natural habitat is changed. These changes have caused many living things to become endangered. Some animals move to new habitats.

Think about your neighborhood. Do you think the environment always looked this way, or has it changed over time?

Roads Replaced Natural Environment

Homes and roads have been built where a natural environment once existed. The plants and animals that used to live here cannot survive as well now that the environment has changed.

Look back at the sentence you highlighted. How is it connected to something you have learned earlier in this lesson?

_____

_____

_____

_____

How Has Your Environment Changed?

Activity 13
Evaluate Like a Scientist

Quick Code: us3566s

Human Impacts

Give an example of a human activity that affects a habitat. **Explain** how the activity has an effect on the habitat by analyzing at least three steps. For example:

- Step 1: human action
- Step 2: local effect on the habitat
- Step 3: effect on the plants or animals within the ecosystem

_____
_____
_____
_____

SEP Obtaining, Evaluating, and Communicating Information
CCC Cause and Effect

Concept 3.3: Environmental Changes

3.3 | Learn — How does the environment change over time?

Activity 14

Investigate Like a Scientist

Quick Code: us3567s

Hands-On Investigation: A Changing Ecosystem

In this investigation, you will use sand, plants, and rocks to model flooding of an ecosystem.

Make a Prediction

How do you think flooding can change an ecosystem?

_____

_____

_____

_____

_____

_____

_____

_____

_____

_____

SEP Analyzing and Interpreting Data
CCC Cause and Effect

What materials do you need? (per group)

- Chart paper
- Sand
- Leaves
- Twigs
- Rocks
- Plastic farm animals
- Strainer
- Water
- Large bin, with lid
- Plastic bottle, $1\frac{1}{2}$ L

What Will You Do?

1. Set up your ecosystem model in the large bin using the materials provided.

2. Draw a picture of your ecosystem. Use the water to flood your ecosystem model.

3. Draw a picture of your flooded ecosystem. Observe how the living and nonliving components of your ecosystem are affected.

Concept 3.3: Environmental Changes

3.3 | Learn   How does the environment change over time?

Ecosystem Before Flooding	Ecosystem After Flooding

Think About the Activity

How can flooding change an ecosystem?

_____

_____

_____

_____

Do you think that all the living things survived after the flooding? Why or why not?

_____
_____
_____

How do you think flooding changed the nonliving parts of the ecosystem?

_____
_____
_____

How do changes to the nonliving parts of the ecosystem affect the living parts?

_____
_____
_____
_____

Can you think of one way that humans change an ecosystem? What could be the outcomes of this change?

_____
_____
_____
_____

3.3 | Learn   How does the environment change over time?

> How Can We Reduce the Changes We Make on an Environment?

Activity 15

Investigate Like a Scientist

Quick Code: us3568s

Hands-On Investigation: It's All Downstream

In this investigation, you will create a model of a watershed that includes the San Diego Bay in California. You will then simulate the introduction of pollutants in the watershed and observe how pollutants can travel downstream and affect many different water resources.

Make a Prediction

What do you think will happen when pollution enters a watershed?

_____

_____

_____

_____

SEP Planning and Carrying Out Investigations

176

What materials do you need? (per group)

- Aluminum foil
- Medium-sized hardcover book
- Vegetable oil
- 2 cups of water
- Aluminum foil pan, 13 × 9 × 2
- Map of California with watersheds clearly delineated
- Modeling clay
- Food coloring

What Will You Do?

1. Create a model of a watershed using the materials provided.
2. Investigate what happens when you add water to the watershed model.
3. Record your observations on the Watershed Model Chart.
4. Investigate the effect of pollution (the colored oil) on the watershed.

Concept 3.3: Environmental Changes

3.3 | Learn How does the environment change over time?

Watershed Model

Trial #	Water Quality	Prediction: Where will the water move?	Observation: What did the water do?	What would be the effect of the water flow?
Trial 1	Clean			
Trial 2	Polluted			

Think About the Activity

Compare the watershed models among the class. What happens when pollution enters a watershed?

_____
_____
_____
_____

What does the phrase "we all live downstream" mean?

_____
_____
_____
_____

Explain why scientists monitor the health and quality of our water resources.

_____
_____
_____
_____
_____
_____
_____
_____

Activity 16
Analyze Like a Scientist

Minimizing Human Impact

Quick Code: us3569s

Read the text and **watch** the videos. As you read and watch, **think** about the ways that humans can reduce their impact on the environment. At the end, **write** a list of your ideas. **Share** your list to create a class list.

Minimizing Human Impact

Predicting the negative effects human activities can have on ecosystems can allow humans to take steps to avoid harming ecosystems. We really can learn from our past! Humans can use scientific knowledge to understand past environmental changes and apply this knowledge to new situations. If we define problems clearly, then we can design solutions to reduce negative effects.

Video — Reduce, Reuse, Recycle

SEP Constructing Explanations and Designing Solutions

180

We know that cutting down trees destroys habitats and causes erosion. If a **community** needs to construct new buildings, it could build in areas that are already paved instead of disrupting forests. We

Canada: Nature Takes Over

can all make changes that reduce our negative impact on our ecosystems. Some simple ways are to use less (reduce), reuse items instead of throwing them away, and **recycle** materials.

We can solve some problems by preventing them. For example, governments can pass laws that make it harder for people to pollute the environment. Public education is also important. For example, exhaust from cars pollutes the air. People can learn how to reduce the amount of driving they do.

Make a list of ways that humans can reduce their impact on the environment.

_____

_____

_____

_____

_____

Activity 17

Evaluate Like a Scientist

Quick Code: us3570s

Moving Pollution

Use evidence from your reading, videos, and other activities you have completed to **construct** an argument for the following case:

A person dumps used motor oil on the ground instead of bringing it to a waste facility. That person states that it is okay because the motor oil will only affect a small area where nothing is growing.

Use what you know about the importance of predicting the effects of human activities on ecosystems to construct an argument with a different viewpoint. **Construct** an argument that explains how dumping oil on the ground could affect nearby water ecosystems. **Support** your argument with evidence from a viewed resource or completed activity from this lesson.

- **SEP** Engaging in Argument from Evidence
- **SEP** Constructing Explanations and Designing Solutions
- **CCC** Cause and Effect

Concept 3.3: Environmental Changes

3.3 | Learn — How does the environment change over time?

Write your argument here.

Concept 3.3: Environmental Changes

3.3 | Share — How does the environment change over time?

Activity 18

Record Evidence Like a Scientist

Quick Code: us3571s

The Salt Harvest Mouse

Now that you have learned more about how habitats change over time, go back and complete your KWL Chart from Day 1. Be sure to include information about how the marsh has changed and use this information to make inferences about how the habitat changes have caused the mouse to become endangered.

Look at the image below. Then, **answer** the following questions.

Let's Investigate the Salt Harvest Mouse

Talk Together

How can you describe environmental changes for the salt harvest mouse now? How is your explanation different from before?

SEP Constructing Explanations and Designing Solutions

Look at the Can You Explain? question. You first read this question at the beginning of the lesson.

> **Can You Explain?**
>
> How does the environment change over time?

To plan your scientific explanation, first **write** your claim. Your claim is a one-sentence answer to the question you investigated. It answers the question: *What can you conclude?* It should not start with *yes* or *no*.

My claim:

```
```

Data 1

Data 2

Concept 3.3: Environmental Changes | **187**

3.3 | Share
How does the environment change over time?

Topic: _____

Evidence	How the Evidence Supports the Claim

188

Now, **write** your scientific explanation.

The environment changes over time...

Concept 3.3: Environmental Changes

STEM in Action

Quick Code: us3572s

Activity 19

Analyze Like a Scientist

Careers and Short-Term Changes in Ecosystems

Read the text about naturalists and **watch** the videos. **Discuss** the role that naturalists play in teaching people to care for and protect ecosystems.

Careers and Short-Term Changes in Ecosystems

A scientific understanding of nature helps us to better manage ecosystems. We can help prevent ecosystem damage caused by pollution and development. We can also help restore ecosystems that were once damaged. Protecting and restoring ecosystems requires knowledge.

Video — Naturalists

But it also requires more than scientists. People make choices every day that affect ecosystems. Helping people understand nature and the effects of their choices is one of the most important jobs of a naturalist.

Protecting the Great Barrier Reef

You may have seen some famous naturalists on TV. But did you know that naturalists work in communities everywhere? Many local parks and nature preserves offer programs to help teach people about their local environments. Naturalists design and run these programs to teach and inspire people to take better care of local ecosystems.

Naturalists Today

Construct an argument for why it is important to have naturalists in every community. Use your understanding of ecosystem changes to **support** your argument. Then, **suggest** a way that a naturalist could rely on technology such as digital cameras, video recorders, or computer programs to do their job.

_____
_____
_____
_____
_____
_____
_____
_____
_____
_____
_____
_____

Concept 3.3: Environmental Changes

3.3 | Share — How does the environment change over time?

Activity 20

Evaluate Like a Scientist

Quick Code: us3573s

Review: Environmental Changes

Think about what you have read and seen in this lesson. **Write** some key ideas you have learned. **Review** your notes with a partner. Your teacher may also have you take a practice test.

_____
_____
_____
_____
_____
_____
_____
_____

Talk Together

Think about what you saw in Get Started. Use your new ideas to discuss how animals are affected by environmental changes over time.

Unit Project

Solve Problems Like a Scientist

Unit Project: Environmental Changes and Animals

Quick Code: us3575s

In this project, you will use what you know about how Earth and the environment change to create a story of the geologic history for your area. **Read** the text and **complete** the activities that follow.

Paleontologist

SEP Planning and Carrying Out Investigations
CCC Stability and Change

Paleontologists study ancient rocks to determine the history of an area. They can use a number of tools to determine the age of the rock. They can also use the type of rock to determine what happened in the past. For example, when paleontologists see limestone, they know that that area was covered by warm, clear, shallow water in the past. The animals that they find in this limestone are generally different from the ones you would find in the area now. This is because the environment changes. Fish can live in shallow water, but if the area is currently a desert, no fish can live there.

Did you know that the Mojave Desert has not always been a desert? At times, it has been part of an ancient ocean and was once covered with large lakes. As a result, it has been home to different plants and animals. When it was underwater, animals that can live underwater thrived there. When it was moist and covered with lakes, many animals were able to flourish that could not survive there now. How did the animals change in response to these environmental changes? How have the animals in your area changed throughout history?

Unit Project

Geologic History

Research the geologic history for your area. **Learn** about two different periods in the past. **Take notes** about the plants, animals, and environment in the chart.

Plants	Animals	Environment

Create a story that shows the environment where you live now as well as at two other times in history. **Write** your story on the lines. **Draw** pictures of the plants and animals in the boxes. Be sure to include detailed dates related to the other time periods you picked.

_____

_____

_____

_____

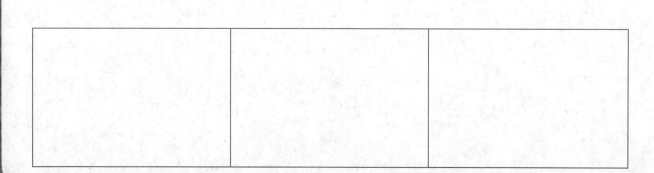

Changing Animals and Plants

For the animals and plants in your story, **explain** why your area was a good environment for these living things at the time. **Write** or **draw** your answers.

Organism	Why the Environment Was Good

Grade 3 Resources

- **Bubble Map**
- **Safety in the Science Classroom**
- **Vocabulary Flash Cards**
- **Glossary**
- **Index**

Name _____

Bubble Map

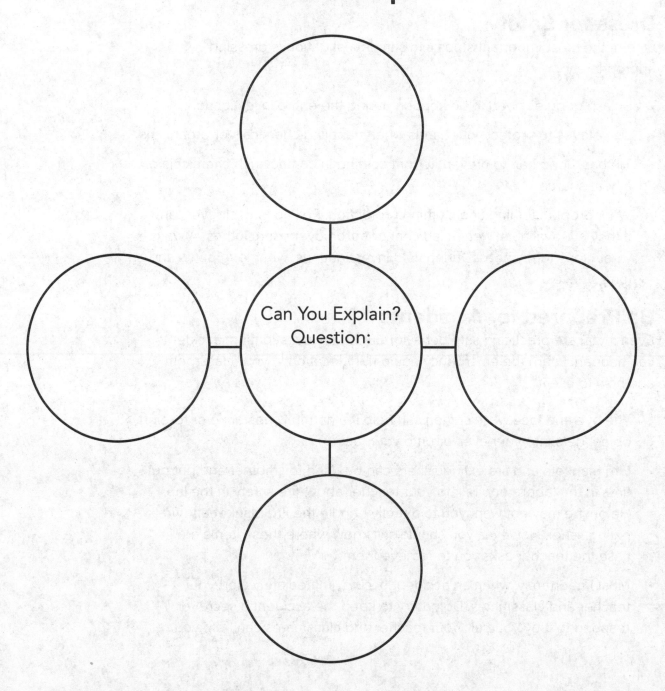

Bubble Map | R3

Safety

Safety in the Science Classroom

Following common safety practices is the first rule of any laboratory or field scientific investigation.

Dress for Safety

One of the most important steps in a safe investigation is dressing appropriately.

- Splash goggles need to be kept on during the entire investigation.
- Use gloves to protect your hands when handling chemicals or organisms.
- Tie back long hair to prevent it from coming in contact with chemicals or a heat source.
- Wear proper clothing and clothing protection. Roll up long sleeves, and if they are available, wear a lab coat or apron over your clothes. Always wear close toed shoes. During field investigations, wear long pants and long sleeves.

Be Prepared for Accidents

Even if you are practicing safe behavior during an investigation, accidents can happen. Learn the emergency equipment location in your classroom and how to use it.

- The eye and face wash station can help if a harmful substance or foreign object gets into your eyes or onto your face.
- Fire blankets and fire extinguishers can be used to smother and put out fires in the laboratory. Talk to your teacher about fire safety in the lab. He or she may not want you to directly handle the fire blanket and fire extinguisher. However, you should still know where these items are in case the teacher asks you to retrieve them.
- Most importantly, when an accident occurs, immediately alert your teacher and classmates. Do not try to keep the accident a secret or respond to it by yourself. Your teacher and classmates can help you.

Practice Safe Behavior

There are many ways to stay safe during a scientific investigation. You should always use safe and appropriate behavior before, during, and after your investigation.

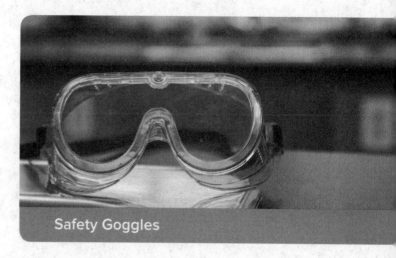
Safety Goggles

- Read the all of the steps of the procedure before beginning your investigation. Make sure you understand all the steps. Ask your teacher for help if you do not understand any part of the procedure.

- Gather all your materials and keep your workstation neat and organized. Label any chemicals you are using.

- During the investigation, be sure to follow the steps of the procedure exactly. Use only directions and materials that have been approved by your teacher.

- Eating and drinking are not allowed during an investigation. If asked to observe the odor of a substance, do so using the correct procedure known as wafting, in which you cup your hand over the container holding the substance and gently wave enough air toward your face to make sense of the smell.

- When performing investigations, stay focused on the steps of the procedure and your behavior during the investigation. During investigations, there are many materials and equipment that can cause injuries.

- Treat animals and plants with respect during an investigation.

- After the investigation is over, appropriately dispose of any chemicals or other materials that you have used. Ask your teacher if you are unsure of how to dispose of anything.

- Make sure that you have returned any extra materials and pieces of equipment to the correct storage space.

- Leave your workstation clean and neat. Wash your hands thoroughly.

Vocabulary Flash Cards

air

the part of the atmosphere closest to Earth; the part of the atmosphere that organisms on Earth use for respiration

ancient

extremely old

Arctic

being from an icy climate, such as the north pole

behavior

the way in which a living thing acts

camouflage

the coloring or patterns on an animal's body that allow it to blend in with its environment

carnivore

a meat eater

community

a group of different populations that live together and interact in an environment

coral reef

a structure formed by the hard skeletons of animals that live in warm, shallow ocean water

desert

an area that gets little precipitation and has very little vegetation

dinosaur

an extinct organism with reptile and birdlike features: Dinosaurs lived on Earth millions of years ago

ecosystem

all the living and nonliving things in an area that interact with each other

energy

the ability to do work or cause change; the ability to move an object some distance

Vocabulary Flash Cards | R11

environment

Image: Odua Images / Shutterstock.com

all the living and nonliving things that surround an organism

extinct

Image: Ocean First Education

when a species is no longer surviving

factor

Image: Paul Fuqua

something that causes another thing to move or change

food web

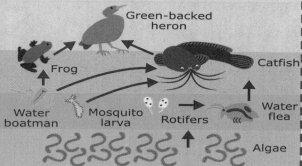

a model that shows many different feeding relationships among living things

Vocabulary Flash Cards | R13

fossil

rocks that have the imprint of living things that were on Earth millions of years ago

grassland

an area of grass covered land that animals use to eat and roam

habitat

the location in which an organism lives

herbivore

a plant eater

Vocabulary Flash Cards | R15

impact

to influence or affect something

instinct

behaviors an animal is born with that help them to survive; inherited behaviors

interact

to act on one another

microorganisms

the tiniest of living things, can only be seen under a microscope

Vocabulary Flash Cards | R17

natural

not human-made

nutrient

important particles found in food that a living thing needs to survive

organism

any individual living thing

pollution

when harmful materials have been put into the air, water, or soil

Vocabulary Flash Cards | R19

predator

an animal that hunts and eats another animal

prehistoric

from a time before human history was recorded

recycle

to create new materials from used products

reproduce

to make more of a species; to have offspring

species

Image: Paul Fuqua

a group of the same kinds of living things

survival

Image: Discovery Communications, Inc.

ability to live and remain alive

survive

Image: Paul Fuqua

to continue living or existing: An organism survives until it dies; a species survives until it becomes extinct.

temperature

Image: Discovery Communications, Inc.

a measure of how hot or cold a substance is

trait

a characteristic or property of an organism

tropical

being from a warmer climate especially near the equator

water

a compound made of hydrogen and oxygen; can be in either a liquid, ice, or vapor form and has no taste or smell

Vocabulary Flash Cards | R25

Glossary

English ——— A ——— Español

adapt
something a plant or animal does to help it survive in its environment

adaptarse
algo que una planta o animal hace para sobrevivir en su medio ambiente

adaptation
how a plant or animal has changed over time to help it survive in its environment (related word: adapt)

adaptación
manera en la que ha cambiado una planta o un animal con el transcurso del tiempo para sobrevivir en su medio ambiente (palabra relacionada: adaptar)

adjust
to change one's position or behavior to allow for a better fit, to adapt

acomodarse
cambiar de posición o comportamiento para ajustarse mejor, adaptarse

air
a gas that is all around you and you can't see it, but living things like plants and animals need it to breathe and to grow

aire
gas que nos rodea y que no podemos ver, pero que las plantas y los animales necesitan para respirar y crecer

air pressure
the force that air puts on an area (related word: pressure)

presión de aire
fuerza que el aire ejerce sobre un área (palabra relacionada: presión)

analyze
to closely examine something and then explain it

analizar
examinar con atención algo y luego explicarlo

ancient
very old

antiguo
extremadamente viejo

arctic
being from an icy climate, such as the North Pole

ártico
que pertenece a un clima helado, como el Polo Norte

artificial selection
specifically breeding animals or cultivating plants only for certain desired genetic outcomes

selección artificial
criar animales o cultivar plantas específicamente para obtener determinados resultados genéticos deseados

atmosphere
layers of gas that surround a planet (related word: atmospheric)

atmósfera
capas de gas que rodean un planeta (palabra relacionada: atmosférico)

attract
to pull one thing toward another (related word: attraction)

atraer
jalar un objeto hacia otro (palabra relacionada: atracción)

―― B ――

balanced forces
when two equal forces are applied to an object in opposite directions, the object does not move

fuerza equilibrada
cuando se aplican dos fuerzas iguales sobre un objeto en direcciones opuestas, el objeto no se mueve

barometer
a tool used to measure air pressure (related word: barometric)

barómetro
herramienta usada para medir la presión del aire (palabra relacionada: barométrico)

behavior
the way in which a living thing acts (related word: behave)

conducta
manera en la que actúa un ser vivo (palabra relacionada: comportarse)

— C —

camouflage
the coloring or patterns on an animal's body that allow it to blend in with its environment

camuflaje
color o patrones del cuerpo de un animal que le permite mezclarse con su medioambiente

carnivore
a meat eater

carnívoro
que se alimenta de carne

characteristic
a special quality that something may have

característica
cualidad especial de algo

climate
the usual weather conditions in a place or area (related word: climatic)

clima
condiciones del tiempo atmosférico habituales en un lugar o área (palabra relacionada: climático)

coast

an area where the ocean meets the land

costa

área donde el océano se encuentra con la tierra

---

community

a group of different populations that live together and interact in an environment

comunidad

grupo de distintas poblaciones que viven juntas e interactúan en un ambiente

---

contact

when two things are so close they touch

contacto

cuando dos objetos están tan cerca que se tocan

---

coral reef

an area that forms in the warm, shallow ocean waters made from the hard skeletons of animals called corals

arrecife de coral

área que se forma en aguas marinas cálidas y poco profundas a partir del esqueleto duro de animales llamados corales

---

cycle

a process that repeats (related word: cyclical)

ciclo

proceso que se repite (palabra relacionada: cíclico)

D

data
measurements or observations (related word: datum)

datos
medidas u observaciones (palabra relacionada: dato)

desert
an area that gets very little rain water and does not have a lot of growing plants

desierto
área que recibe muy poca precipitación y tiene muy poca vegetación

detect
to notice or find, often with the help of a science tool (related words: detection, detector)

detectar
notar o encontrar, generalmente con la ayuda de una herramienta científica (palabras relacionadas: detección, detector)

dinosaur
an extinct organism with reptile and birdlike features: Dinosaurs lived on Earth millions of years ago

dinosaurio
organismo extinto con características de reptil y ave: los dinosaurios vivían en la Tierra hace millones de años

discharge
the release of energy

descarga
liberación de energía

drought
a long period of little or no rain

sequía
escasez o ausencia prolongada de lluvia

--- E ---

ecosystem
all the living and nonliving things in an area that interact with each other

ecosistema
todos los seres vivos y objetos sin vida de un área, que se interrelacionan entre sí

electrical charges
a type of charge, either positive, negative, or neutral

carga eléctrica
un tipo de carga, ya sea positiva, negativa, o neutra

electrical energy
energy produced by power plants that flows through electrical lines and wires

energía eléctrica
energía producida por centrales eléctricas que fluye a través de cables y líneas eléctricas

electromagnet
a metal object that acts as a magnet when an electric current moves through it

electroimán
objeto de metal que actúa como un imán cuando una corriente eléctrica pasa a través de él

endangered
a type of plant or animal that is in danger of becoming extinct

amenazado
tipo de planta o animal que está en peligro de extinción

energy
the ability to do work or make something change

energía
capacidad para hacer un trabajo o producir un cambio

environment
all the living and nonliving things that surround an organism

medio ambiente
todos los seres vivos y objetos sin vida que rodean a un organismo

equator
an imaginary line that divides Earth into Northern and Southern Hemispheres; located halfway between the North and South Poles (related word: equatorial)

ecuador
línea imaginaria que divide la Tierra en Hemisferio Norte y Hemisferio Sur; ubicada a mitad de camino entre el Polo Norte y el Polo Sur (palabra relacionada: ecuatorial)

evidence
facts that give us more information, clues, or proof about something else

evidencia
hechos que nos dan más información, pistas o pruebas sobre otra cosa

extinct
when a plant or an animal is no longer in existance (related word: extinction)

extinto
cuando una planta o un animal ya no existe (palabra relacionada: extinción)

F

factor
something that influences another thing to move or change

factor
algo que influye en que otra cosa se mueva o cambie

food web
a model that shows many different feeding relationships among living things

red alimentaria
modelo que muestra muchas y diferentes relaciones de alimentación entre los seres vivos

Glossary | R35

force
a pull or push that is applied to an object

fuerza
acción de atraer o empujar que se aplica a un objeto

forecast
(v) to analyze weather data and make an educated guess about weather in the future; (n) a prediction about what the weather will be like in the future based on weather data

pronosticar / pronóstico
(v) analizar los datos del tiempo y hacer una conjetura informada sobre el tiempo en el futuro; (s) predicción sobre cómo será el tiempo en el futuro en base a datos

fossil
the remains of a living animal or plant from a very long time ago (related word: fossilize)

fósil
restos de un animal o planta de hace mucho tiempo (palabra relacionada: fosilizar)

friction
when two objects rub against each other

fricción
cuando dos objetos se frotan entre sí

G

generation
the next group of living things or species that will be born around the same time

generación
el siguiente grupo de seres vivos o especies que nacerán alrededor de la misma época

germination
when a young plant grows from a seed

germinación
proceso por el cual una planta joven brota de una semilla (palabra relacionada: germinar)

grassland
a large area of land covered by grass

pradera
área de tierra cubierta principalmente de hierba

gravity
the force that pulls an object toward the center of Earth (related word: gravitational)

gravedad
fuerza que jala a un objeto hacia el centro de la Tierra (palabra relacionada: gravitacional)

H

habitat
the place where a plant or animal lives

hábitat
lugar donde vive una planta o un animal

heat
a form of energy; the state of being very warm

calor
forma de energía; estado de tener una temperatura muy alta

herbivore
a plant eater

herbívoro
que se alimenta de vegetales

humidity
the measure of how much water vapor is in the air

humedad
medida de cuánto vapor de agua hay en el aire

hurricane
a storm with strong winds and rain that forms over tropical waters (related terms: typhoon, tropical cyclone)

huracán
tormenta con fuertes vientos y lluvia que se forma sobre aguas tropicales (palabras relacionadas: tifón, ciclón tropical)

I

impact
to influence or affect something

impactar
afectar o influir en algo

inherit
to get genetic information and traits from a parent or parents (related word: inheritance)

heredar
obtener información y rasgos genéticos de uno o ambos padres (palabra relacionada: herencia)

instinct
behaviors animals and people are born with that help them survive

instinto
conductas con las que nacen los animales y las personas y que los ayudan a sobrevivir

interact
to act on one another (related word: interaction)

interactuar
ejercer influencia mutua (palabra relacionada: interacción)

L

life cycle
the various stages of an organism's development and reproduction

ciclo de la vida
diversas etapas del desarrollo y de la reproducción de un organismo

lifespan
how long in time an organism is expected to live

longevidad
cuánto tiempo se espera que viva un organismo

lightning
when electricity flows between a cloud and the ground or between two clouds and you sometimes see a streak or a flash in the sky

relámpago
cuando fluye electricidad entre una nube y el suelo o entre dos nubes, y a veces se ve una raya o un destello en el cielo

--- M ---

magnetic
having the properties of a magnet; having the ability to be attracted to or by a magnet

magnético
que tiene las propiedades de un imán; que tiene la capacidad de ser atraído hacia o por un imán

magnetic field
a region in space near a magnet or electric current in which magnetic forces can be detected

campo magnético
región en el espacio cerca de un imán o de una corriente eléctrica, donde pueden detectarse fuerzas magnéticas

magnetism
the amount of attraction to a magnet

magnetismo
la cantidad de atracción hacia un imán

mature
when a living thing is fully grown or an adult (related word: maturity)

maduro / madurar
organismo que ha crecido por completo, o que es adulto (palabra relacionada: madurez)

metamorphosis
when a living thing goes through changes during its life cycle, like a frog or a butterfly

metamorfosis
cuando un ser vivo experimenta cambios durante su ciclo de vida, como en el caso de las ranas o las mariposas

meteorology
the study of patterns of weather

meteorología
estudio de los patrones del tiempo atmosférico

microorganisms
the tiniest of living things, can only be seen under a microscope

microorganismo
los seres vivos más diminutos, que sólo se pueden ver con un microscopio

migrating
traveling within a group to a different location during season changes

migratorio
que viaja dentro de un grupo a un lugar diferente durante los cambios de estación

motion
when something moves from one place to another (related terms: move, movement)

movimiento
cuando algo pasa de un lugar a otro (palabras relacionadas: mover, desplazamiento)

— N —

natural
not human-made (related word: nature)

natural
que no está hecho por un ser humano (palabra relacionada: naturaleza)

negative charge
a charge that you get when there is a build-up of electrons

carga negativa
carga que resulta de la acumulación de electrones

neutral
having no electrical charge, being neither positive nor negative

neutro
que no tiene carga eléctrica, que no es positivo ni negativo

nutrient
something in food that helps people, animals, and plants live and grow

nutriente
algo en los alimentos que ayuda a las personas, los animales y las plantas a crecer

observe
to study something using your senses (related word: observation)

observar
estudiar algo usando tus sentidos (palabra relacionada: observación)

offspring
a new organism that is produced by one or more parents

descendencia
organismo nuevo originado por uno o más progenitores

organism
any individual living thing

organismo
todo ser vivo individual

P

parasite
a plant, animal, or fungus that lives on or in another living thing to get food and energy from it

parásito
una planta, animal, u hongo que vive sobre o dentro de otro ser vivo, del cual se alimenta y obtiene energía

pendulum
a string or bar that is loose at one end but fixed at the other end and can swing back and forth, like in a clock

péndulo
cuerda o barra que posee un extremo suelto y otro fijo y puede balancearse de un lado a otro como en un reloj

pole
the opposite ends of a battery, a magnet, or the north and south ends of Earth

polo
extremos opuestos de una batería, un imán, o los extremos norte y sur de Tierra

pollen
the yellow powder found inside of a flower (related word: pollinate)

polen
polvo amarillo que se encuentra dentro de una flor (palabra relacionada: polinizar)

pollution
when harmful materials have been put into the air, water, or soil (related word: pollute)

contaminación
cuando se introducen materiales perjudiciales en el aire, el agua o el suelo (palabra relacionada: contaminar)

positive charge
a charge that you get when there are more protons than electrons

carga positiva
carga que resulta cuando hay más protones que electrones

precipitation
water that is released from clouds in the sky; includes rain, snow, sleet, hail, and freezing rain

precipitación
agua liberada de las nubes en el cielo; incluye la lluvia, la nieve, la aguanieve, el granizo, y la lluvia congelada

predators
the larger animals that hunt the smaller animals, or prey, for food

depredador
animales más grandes que cazan a otros más pequeños, o presas, para alimentarse

predict
to make a guess based on what you already know (related word: prediction)

predecir
hacer una suposición a partir de lo que ya se sabe (palabra relacionada: predicción)

prehistoric
a time before history was written

prehistórico
época antes de que se escribiera la historia

prey
the animals that get hunted by the larger animals, or predators, for food

presa
animales que son cazados por animales más grandes, o depredadores, como alimento

R

rain
liquid water that falls from the sky

lluvia
agua líquida que cae desde el cielo

recycle
to create new materials from used products

reciclar
crear nuevos materiales a partir de productos usados

region
a place, especially around the world

región
lugar, especialmente alrededor del mundo

repel

to force an object away or to keep it away

repeler

forzar a un objeto para que se aleje o mantenerlo alejado

reproduce

to make more of a species; to have offspring (related word: reproduction)

reproducir

engendrar más individuos de una especie; tener descendencia (palabra relacionada: reproducción)

---- S ----

seed

the small part of a flowering plant that grows into a new plant

semilla

parte pequeña de una planta con flor que crece y se convierte en una nueva planta

seedling

a young plant that grows from a seed

plántula

planta joven que crece de una semilla

severe

dangerous or harsh conditions

severo

condiciones peligrosas o adversas

species
a group of the same kinds of living things

especie
un grupo de las mismas clases de seres vivos

static electricity
electric charges that build up on an object

electricidad estática
cargas eléctricas que se acumulan sobre un objeto

stored energy
energy in an object or substance that is not being given off by the object or substance

energía almacenada
energía en un objeto o una sustancia que no es liberada por el objeto o la sustancia

survival
ability to live and remain alive

supervivencia
capacidad de vivir y mantenerse vivo

survive
to continue living or existing: an organism survives until it dies; a species survives until it becomes extinct (related word: survival)

sobrevivir
continuar viviendo o existiendo: un organismo sobrevive hasta que muere; una especie sobrevive hasta que se extingue (palabra relacionada: supervivencia)

T

temperature (general)
a measure of how hot or cold a substance is

temperatura (general)
medida de cuán caliente o fría es una sustancia

tornado
a funnel-shaped cloud or column of air that rotates at high speeds and extends downward from a cloud to the ground

tornado
nube o columna de aire con forma de embudo que rota a altas velocidades y se extiende hacia abajo desde una nube hasta el suelo

trait
a characteristic that you get from one of your parents

rasgo
característica que se obtiene de uno de los progenitores

tropical
from a warmer climate, near the equator

tropical
que pertenece a un clima más cálido, cerca del ecuador

W

water
a clear liquid that has no taste or smell

agua
líquido transparente que no tiene sabor ni olor

weather
the properties of the atmosphere at a given time and location, including temperature, air movement, and precipitation

tiempo atmosférico
propiedades de la atmósfera en un determinado momento y lugar; entre ellas, la temperatura, el movimiento del aire y las precipitaciones

wind
the movement of air due to atmospheric pressure differences

viento
movimiento de aire que se produce por las diferencias en la presión atmosférica

work
a force applied to an object over a distance

trabajo
fuerza aplicada a un objeto a lo largo de una distancia

Index

A

Adjust 142
Air 82
Analyze Like a Scientist 13, 22, 25–27, 36–37, 49–51, 80–84, 91–95, 98–102, 111–118, 131–133, 142–144, 151–153, 168–170, 180–182, 190–192
Ancient 8, 22, 36–37, 197
Arctic 37, 113, 114
Ask Questions Like a Scientist 10–12, 58–60, 140–141

B

Behavior 99–101, 115

C

Camouflage 112
Can You Explain? 8, 47–48, 56, 127, 138, 187
Community 181, 192
Coral reef 112

D

Desert 111, 115–116, 197
Dinosaur 12, 22

E

Ecosystems
   long-term changes in 148–152, 180–181
   short-term changes in 161–162, 172–175, 190–191
Energy 82
Environments
   changes to 151–152, 169, 197
   human impact on 180–181
   impacting other organisms 92, 100–101, 142–144
   in the past 12, 36, 80,
Evaluate Like a Scientist 18, 38–39, 42–45, 52–53, 68–72, 103, 122–125, 134–135, 145–146, 158, 171, 183–184
Evidence 12
Extinct 25, 44, 49, 132

F

Factors 100
Fossils
   information from 25–31, 36–37, 42–45, 49–50
   properties of 12, 16–17, 22

H

Habitats
  characteristics of 68–72, 78–83, 131–133
  organisms in 85–90, 92, 111–112
  types of 104–106, 119–125
  changes to 142–144, 152, 181
Hands-On Activities 32–34, 64–67, 172–179

I

Impact
  human 168, 180–182
  on habitats 142, 154
Instinct 99
Interact 100
Investigate Like a Scientist 32–34, 64–67, 172–179

M

Microorganism 22

N

Natural 142
Nutrient 91

O

Observe Like a Scientist 14–17, 23–24, 35, 40–41, 61–63, 74–79, 96–97, 104–110, 148–150, 154–157, 159–162
Organisms
  in habitats 100, 111–169
  changes in 25
  traits of 22, 36, 44–45

P

Pollution 152, 176–178, 183
Predator 112, 116
Prehistoric 12

R

Record Evidence Like a Scientist 46–48, 126–130, 186–189
Recycle 181
Reproduce 68, 82

S

Solve Problems Like a Scientist 4–5, 196–199
Species 49

Index

STEM in Action 49–51, 131–133, 192–193
Survival 142
Survive 80, 96–97, 168

T

Temperature 114, 152
Think Like a Scientist 28–31, 85–90, 119–121, 163–167

Trait 36, 101, 116
Tropical 37

U

Unit Project 4–5, 196–199

W

Water 82, 104, 116, 152